ILTS 108 Science: Earth and Space Science

Teacher Certification Exam

By: Sharon Wynne, M.S.

XAMonline, INC.
Boston

Copyright © 2013 XAMonline, Inc.
All rights reserved. No part of the material protected by this copyright notice may be reproduced or utilized in any form or by any means, electronic or mechanical, including photocopying, recording or by any information storage and retrievable system, without written permission from the copyright holder.

To obtain permission(s) to use the material from this work for any purpose including workshops or seminars, please submit a written request to:

XAMonline, Inc.
25 First Street, Suite 106
Cambridge, MA 02141
Toll Free 1-800-301-4647
Email: info@xamonline.com
Web www.xamonline.com

Library of Congress Cataloging-in-Publication Data

Wynne, Sharon A.
Science: Earth and Space Science 108: Teacher Certification / Sharon A. Wynne. -2nd ed.
ISBN 978-1-58197-673-1
1. Science: Earth and Space Science 108   2. Study Guides   3. ILTS
4. Teachers' Certification & Licensure   5. Careers

Disclaimer:
The opinions expressed in this publication are the sole works of XAMonline and were created independently from the Pearson Corporation, National Education Association, Educational Testing Service, or any State Department of Education, National Evaluation Systems or other testing affiliates.

Between the time of publication and printing, state specific standards as well as testing formats and website information may change that is not included in part or in whole within this product. Sample test questions are developed by XAMonline and reflect similar content as on real tests; however, they are not former tests. XAMonline assembles content that aligns with state standards but makes no claims nor guarantees teacher candidates a passing score. Numerical scores are determined by testing companies such as NES or ETS and then are compared with individual state standards. A passing score varies from state to state.

Printed in the United States of America   œ-1
ILTS: Science: Earth and Space Science 108
ISBN: 978-1-58197-673-1

TEACHER CERTIFICATION STUDY GUIDE

Table of Contents

SUBAREA I.   **SCIENCE AND TECHNOLOGY**

COMPETENCY 1.0 UNDERSTAND AND APPLY KNOWLEDGE OF SCIENCE AS INQUIRY 1

Skill 1.1   Recognize the assumptions, processes, purposes, requirements, and tools of scientific inquiry 1

Skill 1.2   Use evidence and logic in developing proposed explanations that address scientific questions and hypotheses 2

Skill 1.3   Identify various approaches to conducting scientific investigations and their applications 3

Skill 1.4   Use tools and mathematical and statistical methods for collecting, managing, analyzing, and communicating results of investigations 4

Skill 1.5   Demonstrate knowledge of ways to report, display, and defend the results of an investigation 6

COMPETENCY 2.0 UNDERSTAND AND APPLY KNOWLEDGE OF THE CONCEPTS, PRINCIPLES, AND PROCESSES OF TECHNOLOGICAL DESIGN 7

Skill 2.1   Recognize the capabilities, limitations, and implications of technology and technological design and redesign 7

Skill 2.2   Identify real-world problems or needs to be solved through technological design 7

Skill 2.3   Apply a technological design process to a given problem situation 7

Skill 2.4   Identify a design problem and propose possible solutions, considering such constraints as tools, materials, time, costs, and laws of nature 8

Skill 2.5   Evaluate various solutions to a design problem 8

EARTH & SPACE SCIENCE

TEACHER CERTIFICATION STUDY GUIDE

COMPETENCY 3.0 UNDERSTAND AND APPLY KNOWLEDGE OF ACCEPTED PRACTICES OF SCIENCE 9

Skill 3.1 Demonstrate an understanding of the nature of science and recognize how scientific knowledge and explanations change over time.. 9

Skill 3.2 Compare scientific hypotheses, predictions, laws, theories, and principles and recognize how they are developed and tested 11

Skill 3.3 Recognize examples of valid and biased thinking in the reporting of scientific research.. 12

Skill 3.4 Recognize the basis for and application of safety practices and regulations in the study of science.. 12

COMPETENCY 4.0 UNDERSTAND AND APPLY KNOWLEDGE OF THE INTERACTIONS AMONG SCIENCE, TECHNOLOGY, AND SOCIETY 15

Skill 4.1 Recognize the historical and contemporary development of major scientific ideas and technological innovations 15

Skill 4.2 Demonstrate an understanding of the ways that science and technology affect people's everyday lives, societal values and systems, the environment, and new knowledge 16

Skill 4.3 Analyze the processes of scientific and technological breakthroughs and their effects on other fields of study, careers, and job markets.. 17

Skill 4.4 Analyze issues related to science and technology at the local, state, national, and global levels 17

Skill 4.5 Evaluate the credibility of scientific claims made in various forums 18

EARTH & SPACE SCIENCE

TEACHER CERTIFICATION STUDY GUIDE

COMPETENCY 5.0   **UNDERSTAND AND APPLY KNOWLEDGE OF THE MAJOR UNIFYING CONCEPTS OF ALL SCIENCES AND HOW THESE CONCEPTS RELATE TO OTHER DISCIPLINES** 19

Skill 5.1   Identify the major unifying concepts of the sciences and their applications in real-life situations 19

Skill 5.2   Recognize connections within and among the traditional scientific disciplines 20

Skill 5.3   Apply fundamental mathematical language, knowledge, and skills at the level of algebra and statistics in scientific contexts 21

Skill 5.4   Recognize the fundamental relationships among the natural sciences and the social sciences 24

SUBAREA II.   **LIFE SCIENCE**

COMPETENCY 6.0   **UNDERSTAND AND APPLY KNOWLEDGE OF CELL STRUCTURE AND FUNCTION** 25

Skill 6.1   Compare and contrast the structures of viruses and prokaryotic and eukaryotic cells 25

Skill 6.2   Identify the structures and functions of cellular organelles 27

Skill 6.3   Describe the processes of the cell cycle 30

Skill 6.4   Explain the functions and applications of the instruments and technologies used to study the life sciences at the molecular and cellular level 34

TEACHER CERTIFICATION STUDY GUIDE

COMPETENCY 7.0 UNDERSTAND AND APPLY KNOWLEDGE OF THE PRINCIPLES OF HEREDITY AND BIOLOGICAL EVOLUTION 35

Skill 7.1 Recognize the nature and function of the gene, with emphasis on the molecular basis of inheritance and gene expression 35

Skill 7.2 Analyze the transmission of genetic information 36

Skill 7.3 Analyze the processes of change at the microscopic and macroscopic levels 38

Skill 7.4 Identify scientific evidence from various sources, such as the fossil record, comparative anatomy, and biochemical similarities, to demonstrate knowledge of theories about processes of biological evolution 39

COMPETENCY 8.0 UNDERSTAND AND APPLY KNOWLEDGE OF THE CHARACTERISTICS AND LIFE FUNCTIONS OF ORGANISMS 42

Skill 8.1 Identify the levels of organization of various types of organisms and the structures and functions of cells, tissues, organs, and organ systems 42

Skill 8.2 Analyze the strategies and adaptations used by organisms to obtain the basic requirements of life 42

Skill 8.3 Analyze factors that influence homeostasis within an organism 44

Skill 8.4 Demonstrate an understanding of the human as a living organism with life functions comparable to those of other life forms 44

TEACHER CERTIFICATION STUDY GUIDE

COMPETENCY 9.0 UNDERSTAND AND APPLY KNOWLEDGE OF HOW ORGANISMS INTERACT WITH EACH OTHER AND WITH THEIR ENVIRONMENT 46

Skill 9.1 Identify living and nonliving components of the environment and how they interact with one another 46

Skill 9.2 Recognize the concepts of populations, communities, ecosystems, and ecoregions and the role of biodiversity in living systems 46

Skill 9.3 Analyze factors that influence interrelationships among organisms 48

Skill 9.4 Develop a model or explanation that shows the relationships among organisms in the environment 49

Skill 9.5 Recognize the dynamic nature of the environment, including how communities, ecosystems, and ecoregions change over time 50

Skill 9.6 Analyze interactions of humans with their environment 51

Skill 9.7 Explain the functions and applications of the instruments and technologies used to study the life sciences at the organism and ecosystem levels 52

SUBAREA III.   PHYSICAL SCIENCE

COMPETENCY 10.0 UNDERSTAND AND APPLY KNOWLEDGE OF THE NATURE AND PROPERTIES OF ENERGY IN ITS VARIOUS FORMS 53

Skill 10.1 Describe the characteristics of and relationships among thermal, acoustical, radiant, electrical, chemical, mechanical, and nuclear energies through conceptual questions 53

Skill 10.2 Analyze the processes by which energy is exchanged or transformed through conceptual questions 53

Skill 10.3 Apply the three laws of thermodynamics to explain energy transformations, including basic algebraic problem solving 54

Skill 10.4 Apply the principle of conservation as it applies to energy through conceptual questions and solving basic algebraic problems 55

EARTH & SPACE SCIENCE

TEACHER CERTIFICATION STUDY GUIDE

COMPETENCY 11.0   UNDERSTAND AND APPLY KNOWLEDGE OF THE STRUCTURE AND PROPERTIES OF MATTER......... 56

Skill 11.1   Describe the nuclear and atomic structure of matter, including the three basic parts of the atom. 56

Skill 11.2   Analyze the properties of materials in relation to their chemical or physical structures and evaluate uses of the materials based on their properties 58

Skill 11.3   Apply the principle of conservation as it applies to mass and charge through conceptual questions 60

Skill 11.4   Analyze bonding and chemical, atomic, and nuclear reactions in natural and man-made systems and apply basic stoichiometric principles 61

Skill 11.5   Apply kinetic theory to explain interactions of energy with matter, including conceptual questions on changes in state 62

Skill 11.6   Explain the functions and applications of the instruments and technologies used to study matter and energy 62

COMPETENCY 12.0 UNDERSTAND AND APPLY KNOWLEDGE OF FORCES AND MOTION 63

Skill 12.1 Demonstrate an understanding of the concepts and interrelationships of position, time, velocity, and acceleration through conceptual questions, algebra-based kinematics, and graphical analysis 63

Skill 12.2 Demonstrate an understanding of the concepts and interrelationships of force, inertia, work, power, energy, and momentum 63

Skill 12.3 Describe and predict the motions of bodies in one and two dimensions in inertial and accelerated frames of reference in a physical system, including projectile motion but excluding circular motion 66

Skill 12.4 Analyze and predict motions and interactions of bodies involving forces within the context of conservation of energy and/or momentum through conceptual questions and algebra-based problem solving 69

Skill 12.5 Describe the effects of gravitational and nuclear forces in real-life situations through conceptual questions 71

Skill 12.6 Explain the functions and applications of the instruments and technologies used to study force and motion in everyday life 72

COMPETENCY 13.0 UNDERSTAND AND APPLY KNOWLEDGE OF ELECTRICITY, MAGNETISM, AND WAVES 73

Skill 13.1 Recognize the nature and properties of electricity and magnetism, including static charge, moving charge, basic RC circuits, fields, conductors, and insulators 73

Skill 13.2 Recognize the nature and properties of mechanical and electromagnetic waves 76

Skill 13.3 Describe the effects and applications of electromagnetic forces in real-life situations, including electric power generation, circuit breakers, and brownouts 77

Skill 13.4 Analyze and predict the behavior of mechanical and electromagnetic waves under varying physical conditions, including basic optics, color, ray diagrams, and shadows 78

TEACHER CERTIFICATION STUDY GUIDE

SUBAREA IV.   **EARTH SYSTEMS AND THE UNIVERSE**

COMPETENCY 14.0 UNDERSTAND AND APPLY KNOWLEDGE OF EARTH'S LAND, WATER, AND ATMOSPHERIC SYSTEMS AND THE HISTORY OF EARTH 81

Skill 14.1   Identify the structure and composition of Earth's land, water, and atmospheric systems and how they affect weather, erosion, fresh water, and soil 81

Skill 14.2   Recognize the scope of geologic time and the continuing physical changes of Earth through time 82

Skill 14.3   Evaluate scientific theories about Earth's origin and history and how these theories explain contemporary living systems 83

Skill 14.4   Recognize the interrelationships between living organisms and Earth's resources and evaluate the uses of Earth's resources 85

COMPETENCY 15.0 UNDERSTAND AND APPLY KNOWLEDGE OF THE DYNAMIC NATURE OF EARTH 87

Skill 15.1   Analyze and explain large-scale dynamic forces, events, and processes that affect Earth's land, water, and atmospheric systems, including conceptual questions about plate tectonics, El Niño, drought, and climatic shifts 87

Skill 15.2   Identify and explain Earth processes and cycles and cite examples in real-life situations, including conceptual questions on rock cycles, volcanism, and plate tectonics 89

Skill 15.3   Analyze the transfer of energy within and among Earth's land, water, and atmospheric systems, including the identification of energy sources of volcanoes, hurricanes, thunderstorms, and tornadoes 93

Skill 15.4   Explain the functions and applications of the instruments and technologies used to study the earth sciences, including seismographs, barometers, and satellite systems 95

COMPETENCY 16.0 UNDERSTAND AND APPLY KNOWLEDGE OF OBJECTS IN THE UNIVERSE AND THEIR DYNAMIC INTERACTIONS 96

Skill 16.1   Describe and explain the relative and apparent motions of the sun, the moon, stars, and planets in the sky 96

Skill 16.2   Recognize properties of objects within the solar system and their dynamic interactions 96

Skill 16.3   Recognize the types, properties, and dynamics of objects external to the solar system 97

COMPETENCY 17.0 UNDERSTAND AND APPLY KNOWLEDGE OF THE ORIGINS OF AND CHANGES IN THE UNIVERSE..... 98

Skill 17.1   Identify scientific theories dealing with the origin of the universe 98

Skill 17.2   Analyze evidence relating to the origin and physical evolution of the universe 98

Skill 17.3   Compare the physical and chemical processes involved in the life cycles of objects within galaxies 99

Skill 17.4   Explain the functions and applications of the instruments, technologies, and tools used in the study of the space sciences, including the relative advantages and disadvantages of earth-based versus space-based instruments and optical versus non-optical instruments 100

TEACHER CERTIFICATION STUDY GUIDE

SUBAREA V. THE EARTH AND ATMOSPHERE

COMPETENCY 18.0 UNDERSTAND AND APPLY KNOWLEDGE OF THE GEOLOGIC PROCESSES AND STRUCTURE OF EARTH 101

Skill 18.1 Relate the dynamic processes that occur in Earth's interior to their causes and effects 101

Skill 18.2 Identify the basic principles of plate tectonic theory, supporting evidence for the theory, and the mechanisms of plate dynamics 102

Skill 18.3 Relate the features and landforms of Earth's surface to the processes that shape them 104

Skill 18.4 Use a geologic column or block diagram to interpret the geologic history of a particular area 113

Skill 18.5 Demonstrate and explain strategies that are used to identify and classify rocks and minerals 116

Skill 18.6 Identify and describe characteristics of various rock types and Minerals and how they are formed 118

COMPETENCY 19.0 UNDERSTAND AND APPLY KNOWLEDGE OF THE HISTORICAL EVOLUTION OF EARTH'S FEATURES AND LIFE FORMS THROUGH GEOLOGIC TIME 121

Skill 19.1 Recognize the scope of geologic time by distinguishing between the human time scale and the geologic time scale 121

Skill 19.2 Identify methods and technologies used to study Earth's history leading to the divisions of geologic time 121

Skill 19.3 Analyze evidence for hypotheses and theories regarding Earth's origins and the evolution of Earth's features and life forms 126

Skill 19.4 Describe how rock strata and fossils lead to inferences about geological environments and climatic conditions 129

COMPETENCY 20.0	UNDERSTAND AND APPLY KNOWLEDGE OF THE TRANSFER OF ENERGY AND CYCLING OF ELEMENTS AND COMPOUNDS IN EARTH SYSTEMS 131
Skill 20.1	Identify chemical reservoirs for various elements and compounds 131
Skill 20.2	Trace the cycle of various elements and compounds through the lithosphere, hydrosphere, biosphere, and atmosphere 132
Skill 20.3	Analyze the factors and interpret data that indicate the transfer of energy within and among Earth's land, water, and atmospheric systems 134
Skill 20.4	Describe the interrelationships among land, water, and atmospheric systems and how these relationships explain various natural processes, cycles, and events 136
Skill 20.5	Identify factors, including human activities, that affect the cycling of elements and compounds in Earth systems 137
COMPETENCY 21.0	UNDERSTAND AND APPLY KNOWLEDGE OF THE STRUCTURE AND PROCESSES OF THE HYDROSPHERE AND ATMOSPHERE 139
Skill 21.1	Describe the physical and chemical nature of water and its influence on the lithosphere, hydrosphere, biosphere, and atmosphere 139
Skill 21.2	Demonstrate knowledge of the characteristics of oceans and freshwater systems 140
Skill 21.3	Analyze atmospheric data and how it correlates to weather patterns, climatic conditions, severe weather, and cloud development 143
Skill 21.4	Explain historic and current technologies and tools associated with data collection and interpretation of meteorologic and climatologic research and predictions 146

COMPETENCY 22.0		UNDERSTAND AND APPLY KNOWLEDGE OF INTERACTIONS BETWEEN HUMANS AND EARTH SYSTEMS 149
Skill 22.1		Identify the types and characteristics of Earth's renewable and nonrenewable resources and the methods employed to preserve or conserve them 149
Skill 22.2		Recognize the effects of human activities on the whole earth 150
Skill 22.3		Analyze how the use of natural resources affects human society economically, socially, and environmentally 151
Skill 22.4		Recognize the effects of Earth processes on human societies through time 152
Skill 22.5		Analyze the use, quality, and scarcity of global freshwater resources, including protection and conservation measures 153

SUBAREA V.   ASTRONOMY

COMPETENCY 23.0		UNDERSTAND AND APPLY KNOWLEDGE OF THE CHARACTERISTICS AND FORMATION OF THE PLANETS AND SOLAR SYSTEM 155
Skill 23.1		Analyze the effects of gravitational force in the solar system 155
Skill 23.2		Identify the general characteristics of the sun, the planets, and their satellites 156
Skill 23.3		Identify the characteristics and orbital nature of comets, asteroids, and meteoroids 161
Skill 23.4		Recognize the scientific basis for understanding various atmospheric, solar, and celestial phenomena, meteors, and auroras 163
Skill 23.5		Recognize historical models of the solar system and the historical development of understanding about the solar system 165

TEACHER CERTIFICATION STUDY GUIDE

COMPETENCY 24.0 UNDERSTAND AND APPLY KNOWLEDGE OF STELLAR EVOLUTION, CHARACTERISTICS OF STARS AND GALAXIES, AND THE FORMATION OF THE UNIVERSE 168

Skill 24.1 Recognize characteristics of various star types, including all the stages and processes of stellar evolution 168

Skill 24.2 Analyze evidence regarding the chemical composition and physical characteristics of stellar and galactic objects 171

Skill 24.3 Evaluate the characteristics and supporting evidence for the various theories of cosmology and cosmogony 173

Skill 24.4 Recognize the historical progression of astronomy and the physical laws that govern it 174

COMPETENCY 25.0 UNDERSTAND AND APPLY KNOWLEDGE OF THE HISTORY AND METHODS OF ASTRONOMY AND SPACE EXPLORATION 175

Skill 25.1 Demonstrate knowledge of the relationship between latitude and the apparent position and motion of celestial objects 175

Skill 25.2 Describe the various technologies used to observe and explore space and the scientific laws that govern space flight 175

Skill 25.3 Recognize the scope and scale of astronomical time and distance 178

Skill 25.4 Recognize the historical technologies used to determine distance and time and their impact on civilization and progress 180

Skill 25.5 Describe the historic progression of space exploration and the technologies that made it possible 181

Sample Test 183

Answer Key 194

Rationales for Sample Questions 195

EARTH & SPACE SCIENCE

TEACHER CERTIFICATION STUDY GUIDE

Great Study and Testing Tips!

What to study in order to prepare for the subject assessments is the focus of this study guide but equally important is *how* you study.

You can increase your chances of truly mastering the information by taking some simple, but effective steps.

Study Tips:

1. Some foods aid the learning process. Foods such as milk, nuts, seeds, rice, and oats help your study efforts by releasing natural memory enhancers called CCKs (*cholecystokinin*) composed of *tryptophan*, *choline*, and *phenylalanine*. All of these chemicals enhance the neurotransmitters associated with memory. Before studying, try a light, protein-rich meal of eggs, turkey, and fish. All of these foods release the memory enhancing chemicals. The better the connections, the more you comprehend.

Likewise, before you take a test, stick to a light snack of energy boosting and relaxing foods. A glass of milk, a piece of fruit, or some peanuts all release various memory-boosting chemicals and help you to relax and focus on the subject at hand.

2. Learn to take great notes. A by-product of our modern culture is that we have grown accustomed to getting our information in short doses (i.e. TV news sound bites or USA Today style newspaper articles.)

Consequently, we've subconsciously trained ourselves to assimilate information better in neat little packages. If your notes are scrawled all over the paper, it fragments the flow of the information. Strive for clarity. Newspapers use a standard format to achieve clarity. Your notes can be much clearer through use of proper formatting. A very effective format is called the *"Cornell Method."*

> Take a sheet of loose-leaf lined notebook paper and draw a line all the way down the paper about 1-2" from the left-hand edge.
>
> Draw another line across the width of the paper about 1-2" up from the bottom. Repeat this process on the reverse side of the page.

Look at the highly effective result. You have ample room for notes, a left hand margin for special emphasis items or inserting supplementary data from the textbook, a large area at the bottom for a brief summary, and a little rectangular space for just about anything you want.

3. Get the concept then the details. Too often we focus on the details and don't gather an understanding of the concept. However, if you simply memorize only dates, places, or names, you may well miss the whole point of the subject.

A key way to understand things is to put them in your own words. If you are working from a textbook, automatically summarize each paragraph in your mind. If you are outlining text, don't simply copy the author's words.

Rephrase them in your own words. You remember your own thoughts and words much better than someone else's, and subconsciously tend to associate the important details to the core concepts.

4. Ask Why? Pull apart written material paragraph by paragraph and don't forget the captions under the illustrations.

Example: If the heading is "Stream Erosion", flip it around to read "Why do streams erode?" Then answer the questions.

If you train your mind to think in a series of questions and answers, not only will you learn more, but it also helps to lessen the test anxiety because you are used to answering questions.

5. Read for reinforcement and future needs. Even if you only have 10 minutes, put your notes or a book in your hand. Your mind is similar to a computer; you have to input data in order to have it processed. *By reading, you are creating the neural connections for future retrieval.* The more times you read something, the more you reinforce the learning of ideas.

Even if you don't fully understand something on the first pass, *your mind stores much of the material for later recall.*

6. Relax to learn so go into exile. Our bodies respond to an inner clock called biorhythms. Burning the midnight oil works well for some people, but not everyone.

If possible, set aside a particular place to study that is free of distractions. Shut off the television, cell phone, and pager and exile your friends and family during your study period.

If you really are bothered by silence, try background music. Light classical music at a low volume has been shown to aid in concentration over other types. Music that evokes pleasant emotions without lyrics is highly suggested. Try just about anything by Mozart. It relaxes you.

7. Use arrows not highlighters. At best, it's difficult to read a page full of yellow, pink, blue, and green streaks. Try staring at a neon sign for a while and you'll soon see that the horde of colors obscure the message.

A quick note, a brief dash of color, an underline, and an arrow pointing to a particular passage is much clearer than a horde of highlighted words.

8. Budget your study time. Although you shouldn't ignore any of the material, *allocate your available study time in the same ratio that topics may appear on the test.*

TEACHER CERTIFICATION STUDY GUIDE

Testing Tips:

1. Get smart, play dumb. Don't read anything into the question. Don't make an assumption that the test writer is looking for something else than what is asked. Stick to the question as written and don't read extra things into it.

2. Read the question and all the choices *twice* before answering the question. You may miss something by not carefully reading, and then re-reading both the question and the answers.

If you really don't have a clue as to the right answer, leave it blank on the first time through. Go on to the other questions, as they may provide a clue as to how to answer the skipped questions.

If later on, you still can't answer the skipped ones . . . *Guess.* The only penalty for guessing is that you *might* get it wrong. Only one thing is certain; if you don't put anything down, you will get it wrong!

3. Turn the question into a statement. Look at the way the questions are worded. The syntax of the question usually provides a clue. Does it seem more familiar as a statement rather than as a question? Does it sound strange?

By turning a question into a statement, you may be able to spot if an answer sounds right, and it may also trigger memories of material you have read.

4. Look for hidden clues. It's actually very difficult to compose multiple-foil (choice) questions without giving away part of the answer in the options presented.

In most multiple-choice questions you can often readily eliminate one or two of the potential answers. This leaves you with only two real possibilities and automatically your odds go to Fifty-Fifty for very little work.

5. Trust your instincts. For every fact that you have read, you subconsciously retain something of that knowledge. On questions that you aren't really certain about, go with your basic instincts. **Your first impression on how to answer a question is usually correct.**

6. Mark your answers directly on the test booklet. Don't bother trying to fill in the optical scan sheet on the first pass through the test.

Just be very careful not to miss-mark your answers when you eventually transcribe them to the scan sheet.

7. Watch the clock! You have a set amount of time to answer the questions. Don't get bogged down trying to answer a single question at the expense of 10 questions you can more readily answer.

TEACHER CERTIFICATION STUDY GUIDE

SUBAREA I. SCIENCE AND TECHNOLOGY

COMPETENCY 1.0 UNDERSTAND AND APPLY KNOWLEDGE OF SCIENCE AS INQUIRY.

Skill 1.1 Recognize the assumptions, processes, purposes, requirements, and tools of scientific inquiry.

Scientific inquiry is an understanding of science through questioning, experimentation and drawing conclusions.

The basic Skills involved in this important process are:

1. Observing
2. Identifying problem
3. Gathering information/research
4. Hypothesizing
5. Experimental design, which includes identifying control, constants, independent and dependent variables
6. Conducting experiment and repeating the experiment for validity
7. Interpreting analyzing and evaluating data
8. Drawing conclusions
9. Communicating conclusions

What are the uses of scientific inquiry?

1. Finding solutions for world problems
2. Encouraging problem solving approach to thinking, learning and understanding
3. To apply math and language Skills
4. To confirm by experimentation that which is already known to the scientific community
5. Offer explanations, conclusions, and critical evaluations
6. Encourage the use of modern technology for research, experiments, analysis, and to communicate data
7. To be up to date with recent advances in science

The simplest form of science inquiry involves the following steps:

1. A question

2. Hypothesis
(A plausible explanation/an educated guess)

3. Experimental design
(Identifying control, constants, independent and dependent variables)

4. Experimenting and repeating the experiment for reliability

5. Data
(Analysis, evaluation and evaluation)

6. Conclusions
Hypothesis correct/incorrect

7. Communicating the conclusions
(Visual, models, written and oral)

Scientific inquiry is a very powerful and highly interesting tool to teach and learn.

Skill 1.2 Use evidence and logic in developing proposed explanations that address scientific questions and hypotheses.

Armed with knowledge of the subject matter, students can effectively conduct investigations. They need to learn to think critically and logically to connect evidence with explanations. This includes deciding what evidence should be used and accounting for unusual data. Based upon data collected during experimentation, basic statistical analysis, and measures of probability can be used to make predictions and develop interpretations.

Students should be able to review the data, summarize, and form a logical argument about the cause-and-effect relationships. It is important to differentiate between causes and effects and determine when causality is uncertain.

When developing proposed explanations, the students should be able to express their level of confidence in the proposed explanations and point out possible sources of uncertainty and error. When formulating explanations, it is important to distinguish between error and unanticipated results. Possible sources of error would include assumptions of models and measuring techniques or devices.

With confidence in the proposed explanations, the students need to identify what would be required to reject the proposed explanations. Based upon their experience, they should develop new questions to promote further inquiry.

Skill 1.3 Identify various approaches to conducting scientific investigations and their applications.

Different types of questions and hypotheses require different approaches to conducting scientific investigations. Some investigations involve making models; some involve discovery of new phenomena and objects; some involve observing and describing objects, organisms, or events; some involve experiments; some involve collecting specimens; and some involve seeking more information.

Different scientific domains use different methods, core theories, and standards.

Scientific investigations sometimes result in new ideas and phenomena to be studied, generate new procedures or methods for an investigation, or develop new technologies to improve data collection. All of these results can lead to new investigations.

Most research in the scientific field is conducted using the scientific method to discover the answer to a scientific problem. The scientific method is the process of thinking through possible solutions to a problem and testing each possibility to find the best solution. The scientific method generally involves the following steps: forming a hypothesis, choosing a method and design, conducting experimentation (collecting data), analyzing data, drawing a conclusion, and reporting the findings. Depending on the hypothesis and data to be collected and analyzed, different types of scientific investigation may be used.

Descriptive studies are often the first form of investigation used in new areas of scientific inquiry. The most important element in descriptive reporting is a specific, clear, and measurable definition of the disease, condition, or factor in question. Descriptive studies always address the five W's: who, what, when, where, and why. They also add an additional "so what?" Descriptive studies include case reports, case-series reports, cross-sectional students, surveillance studies with individuals, and correlational studies with populations. Descriptive studies are used primarily for trend analysis, health-care planning, and hypothesis generation.

A **controlled experiment** is a form of scientific investigation in which one variable, the independent or control variable, is manipulated to reveal the effect on another variable, the dependent (experimental) variable, while are other variables in the system remain fixed. The control group is virtually identical to the dependent variable except for the one aspect whose effect is being tested. Testing the effects of bleach water on a growing plant, the plant receiving bleach water would be the dependent group, while the plant receiving plain water would be the control group. It is good practice to have several replicate samples for the experiment being performed, which allows for results to be averaged or obvious discrepancies to be discarded.

Comparative data analysis is a statistical form of investigation that allows the researcher to gain new or unexpected insight into data based primarily on graphic representation. Comparative data analysis, whether within the research of an individual project or a meta-analysis, allows the researcher to maximize the understanding of the particular data set, uncover underlying structural similarities between research, extract important variables, test underlying assumptions, and detect outliers and anomalies. Most comparative data analysis techniques are graphical in nature with a few quantitative techniques. The use of graphics to compare data allows the researcher to explore the data open-mindedly.

Skill 1.4 **Use tools and mathematical and statistical methods for collecting, managing, analyzing (e.g., average, curve fit, error determination), and communicating results of investigations.**

The procedure used to obtain data is important to the outcome. Experiments consist of **controls** and **variables**. A control is the experiment run under normal conditions. The variable includes a factor that is changed. In biology, the variable may be light, temperature, pH, time, etc. The differences in tested variables may be used to make a prediction or form a hypothesis. Only one variable should be tested at a time. One would not alter both the temperature and pH of the experimental subject.

An **independent variable** is one that is changed or manipulated by the researcher. This could be the amount of light given to a plant or the temperature at which bacteria is grown. The **dependent variable** is that which is influenced by the independent variable.

Average, or arithmetic mean is, the sum of all measurements in the data set divided by the number of observations in the data set.

When one graphs data points, one should follow through by connecting the dots with as smooth a line as possible. The connected points will create either a line or a curve, or appear totally random (you would not be able to connect truly random points). It is possible that not all points will be exactly on the line. Points that do not fall on the line are outliers. If the line is rounded, it is called a **curve fit**. If it is straight, the variables are said to have a linear relationship.

There are many ways in which **errors** could creep in measurements. Errors in measurements could occur because –
1. Improper use of instruments used for measuring – weighing etc.
2. Parallax error – not positioning the eyes during reading of measurements
3. Not using same instruments and methods of measurement during an experiment
4. Not using the same source of materials, resulting in the content of a certain compound used for experimentation

Besides these mentioned above, there could be other possible sources of error as well.

EARTH & SPACE SCIENCE

When erroneous results are used for interpreting data, the conclusions are not reliable. An experiment is valid only when all the constants (like time, place, method of measurement etc.) are strictly controlled. Experimental uncertainty is due to either random errors or systematic errors.

Random errors are defined as statistical fluctuations in the measured data due to the precision limitations of the measurement device. Random errors usually result from the experimenter's inability to take the same measurement in exactly the same way to get exactly the same number.

Systematic errors, by contrast, are defined as reproducible inaccuracies that are consistently in the same direction. Systematic errors are often due to a problem, which persists throughout the entire experiment.

Science uses the **metric system**; as it is accepted worldwide and allows easier comparison among experiments done by scientists around the world.

The meter is the basic metric unit of length. One meter is 1.1 yards. The liter is the basic metric unit of volume. 1 gallon is 3.846 liters. The gram is the basic metric unit of mass. 1000 grams is 2.2 pounds.

The following prefixes are used to describe the multiples of the basic metric units.

deca- 10X the base unit
hecto- 100X the base unit
kilo- 1,000X the base unit
mega- 1,000,000X the base unit
giga- 1,000,000,000X the base unit
tera- 1,000,000,000,000X the base unit

deci - 1/10 the base unit
centi - 1/100 the base unit
milli - 1/1,000 the base unit
micro- 1/1,000,000 the base unit
nano- 1/1,000,000,000 the base unit
pico- 1/1,000,000,000,000 the base unit

The common instrument used for measuring volume is the graduated cylinder. The unit of measurement is usually in milliliters (mL). It is important for accurate measure to read the liquid in the cylinder at the bottom of the meniscus, the curved surface of the liquid.

The common instrument used is measuring mass is the triple beam balance. The triple beam balance is measured in as low as tenths of a gram and can be estimated to the hundredths of a gram.

The ruler or meter sticks are the most commonly used instruments for measuring length. Measurements in science should always be measured in metric units. Be sure when measuring length that the metric units are used.

Skill 1.5 Demonstrate knowledge of ways to report, display, and defend the results of an investigation.

The type of graphic representation used to display observations depends on the data that is collected. **Line graphs** are used to compare different sets of related data or to predict data that has not yet be measured. An example of a line graph would be comparing the rate of activity of different enzymes at varying temperatures. A **bar graph** or **histogram** is used to compare different items and make comparisons based on this data. An example of a bar graph would be comparing the ages of children in a classroom. A **pie chart** is useful when organizing data as part of a whole. A good use for a pie chart would be displaying the percent of time students spend on various after school activities.

As noted before, the independent variable is controlled by the experimenter. This variable is placed on the x-axis (horizontal axis). The dependent variable is influenced by the independent variable and is placed on the y-axis (vertical axis). It is important to choose the appropriate units for labeling the axes. It is best to take the largest value to be plotted and divide it by the number of block, and rounding to the nearest whole number.

Careful research and statistically significant figures will be your best allies should you need to defend your work. For this reason, make sure to use controls, work in a systematic fashion, keep clear records, and have reproducible results.

COMPETENCY 2.0 UNDERSTAND AND APPLY KNOWLEDGE OF THE CONCEPTS, PRINCIPLES, AND PROCESSES OF TECHNOLOGICAL DESIGN.

Skill 2.1 Recognize the capabilities, limitations, and implications of technology and technological design and redesign.

Science and technology are interdependent as advances in technology often lead to new scientific discoveries and new scientific discoveries often lead to new technologies. Scientists use technology to enhance the study of nature and solve problems that nature presents. Technological design is the identification of a problem and the application of scientific knowledge to solve the problem.

While technology and technological design can provide solutions to problems faced by humans, technology must exist within nature and cannot contradict physical or biological principles. In addition, technological solutions are temporary and new technologies typically provide better solutions in the future. Monetary costs, available materials, time, and available tools also limit the scope of technological design and solutions. Finally, technological solutions have intended benefits and unexpected consequences. Scientists must attempt to predict the unintended consequences and minimize any negative impact on nature or society.

Skill 2.2 Identify real-world problems or needs to be solved through technological design.

The problems and needs, ranging from very simple to highly complex, that technological design can solve are nearly limitless. Disposal of toxic waste, routing of rainwater, crop irrigation, and energy creation are but a few examples of real-world problems that scientists address or attempt to address with technology.

Skill 2.3 Apply a technological design process to a given problem situation.

The technological design process has five basic steps:
1. Identify a problem
2. Propose designs and choose between alternative solutions
3. Implement the proposed solution
4. Evaluate the solution and its consequences
5. Report results

After the identification of a problem, the scientist must propose several designs and choose between the alternatives. Scientists often utilize simulations and models in evaluating possible solutions.

Implementation of the chosen solution involves the use of various tools depending on the problem, solution, and technology. Scientists may use both physical tools and objects and computer software.

After implementation of the solution, scientists evaluate the success or failure of the solution against pre-determined criteria. In evaluating the solution, scientists must consider the negative consequences as well as the planned benefits.

Finally, scientists must communicate results in different ways – orally, written, models, diagrams, and demonstrations.

Example:

Problem – toxic waste disposal
Chosen solution – genetically engineered microorganisms to digest waste
Implementation – use genetic engineering technology to create organism capable of converting waste to environmentally safe product
Evaluate – introduce organisms to waste site and measure formation of products and decrease in waste; also evaluate any unintended effects
Report – prepare a written report of results complete with diagrams and figures

Skill 2.4   Identify a design problem and propose possible solutions, considering such constraints as tools, materials, time, costs, and laws of nature.

In addition to finding viable solutions to design problems, scientists must consider such constraints as tools, materials, time, costs, and laws of nature. Effective implementation of a solution requires adequate tools and materials. Scientists cannot apply scientific knowledge without sufficient technology and appropriate materials (e.g. construction materials, software). Technological design solutions always have costs. Scientists must consider monetary costs, time costs, and the unintended effects of possible solutions. Types of unintended consequences of technological design solutions include adverse environmental impact and safety risks. Finally, technology cannot contradict the laws of nature. Technological design solutions must work within the framework of the natural world.

Skill 2.5   Evaluate various solutions to a design problem.

In evaluating and choosing between potential solutions to a design problem, scientists utilize modeling, simulation, and experimentation techniques. Small-scale modeling and simulation help test the effectiveness and unexpected consequences of proposed solutions while limiting the initial costs. Modeling and simulation may also reveal potential problems that scientists can address prior to full-scale implementation of the solution. Experimentation allows for evaluation of proposed solutions in a controlled environment where scientists can manipulate and test specific variables.

COMPETENCY 3.0 UNDERSTAND AND APPLY KNOWLEDGE OF ACCEPTED PRACTICES OF SCIENCE.

Skill 3.1 Demonstrate an understanding of the nature of science (e.g., tentative, replicable, historical, empirical) and recognize how scientific knowledge and explanations change over time.

The combination of science, mathematics and technology forms the scientific endeavor and makes science a success. It is impossible to study science on its own without the support of other disciplines like mathematics, technology, geology, physics, and other disciplines as well.

Science is tentative. By definition it is searching for information by making educated guesses. It must be replicable. Another scientist must be able to achieve the same results under the same conditions at a later time. The term empirical means it must be assessed through tests and observations. Science changes over time. New technologies gather previously unavailable data and enable us to build upon current theories with new information.

Ancient history followed the geocentric theory, which was displaced by the heliocentric theory developed by Copernicus, Ptolemy and Kepler. Newton's laws of motion by Isaac Newton were based on mass, force and acceleration state that the force o gravity between any two objects in the universe depends upon their mass and distance and are still widely used to date. In the 20th century, Albert Einstein was the most outstanding scientist for his work on relativity, which led to his theory that $E=mc2$. Early in the 20th century, Alfred Wegener proposed his theory of continental drift, stating that continents moved away from the super continent, Pangaea. This theory was accepted in 1960s when more evidence was collected on this. John Dalton and Lavosier made significant contributions in the field of atom and matter. The Curies and Ernest Rutherford contributed greatly to radioactivity and the splitting of atom, which have lot of practical applications. Charles Darwin proposed his theory of evolution and Gregor Mendel's experiments on peas helped us to understand heredity. The most significant was the industrial revolution in Britain, in which science was applied practically to increase the productivity and also introduced a number of social problems like child labor.

TEACHER CERTIFICATION STUDY GUIDE

The nature of science mainly consists of three important things:

1. The scientific world view

This includes some very important issues like – it is possible to understand this highly organized world and its complexities with the help of latest technology. Scientific ideas are subject to change. After repeated experiments, a theory is established, but this theory could be changed or supported in future. Only laws that occur naturally do not change.

Scientific knowledge may not be discarded but is modified – e.g., Albert Einstein didn't discard the Newtonian principles but modified them in his theory of relativity.

Scientific knowledge may not be discarded but is modified – e.g., Albert Einstein didn't discard the Newtonian principles but modified them in his theory of relativity.

Also science can't answer all our questions. We can't find answers to questions related to our beliefs, moral values and our norms.

2. Scientific inquiry

Scientific inquiry starts with a simple question. This simple question leads to information gathering, an educated guess otherwise known as hypothesis. To prove the hypothesis, an experiment has to be conducted, which yields data and the conclusion. All experiments must be repeated at least twice to get reliable results. Thus scientific inquiry leads to new knowledge or verifying established theories. Science requires proof or evidence. Science is dependent on accuracy not bias or prejudice. In science, there is no place for preconceived ideas or premeditated results. By using their senses and modern technology, scientists will be able to get reliable information. Science is a combination of logic and imagination. A scientist needs to think and imagine and be able to reason.

Science explains, reasons, and predicts. These three are interwoven and are inseparable. While reasoning is absolutely important for science, there should be no bias or prejudice. Science is not authoritarian because it has been shown that scientific authority can be wrong. No one can determine or make decisions for others on any issue.

3. Scientific enterprise

Science is a complex activity involving various people and places. A scientist may work alone or in a laboratory, classroom or for that matter anywhere. Mostly it is a group activity requiring lot of social Skills of cooperation, communication of results or findings, consultations, discussions etc.

Science demands a high degree of communication to the governments, funding authorities and to public.

EARTH & SPACE SCIENCE

Skill 3.2 Compare scientific hypotheses, predictions, laws, theories, and principles and recognize how they are developed and tested.

Science may be defined as a body of knowledge that is systematically derived from study, observations, and experimentation. Its goal is to identify and establish principles and theories that may be applied to solve problems. Pseudoscience, on the other hand, is a belief that is not warranted. There is no scientific methodology or application. Some of the more classic examples of pseudoscience include witchcraft, alien encounters or any topic that is explained by hearsay.

Scientific theory and experimentation must be repeatable. It is also possible to be disproved and is capable of change. Science depends on communication, agreement, and disagreement among scientists. It is composed of theories, laws, and hypotheses.

theory - the formation of principles or relationships which have been verified and accepted.

law - an explanation of events that occur with uniformity under the same conditions (laws of nature, law of gravitation).

hypothesis - an unproved theory or educated guess followed by research to best explain a phenomena. A theory is a proven hypothesis.

Science is limited by the available technology. An example of this would be the relationship of the discovery of the cell and the invention of the microscope. As our technology improves, more hypotheses will become theories and possibly laws. Science is also limited by the data that is able to be collected. Data may be interpreted differently on different occasions. Science limitations cause explanations to be changeable as new technologies emerge.

The first step in scientific inquiry is posing a question to be answered. Next, a hypothesis is formed to provide a plausible explanation. An experiment is then proposed and performed to test this hypothesis. A comparison between the predicted and observed results is the next step. Conclusions are then formed and it is determined whether the hypothesis is correct or incorrect. If incorrect, the next step is to form a new hypothesis and the process is repeated.

Skill 3.3   Recognize examples of valid and biased thinking in the reporting of scientific research.

Scientific research can be biased in the choice of what data to consider, in the reporting or recording of the data, and/or in how the data are interpreted. The scientist's emphasis may be influenced by his/her nationality, sex, ethnic origin, age, or political convictions. For example, when studying a group of animals, male scientists may focus on the social behavior of the males and typically male characteristics.

Although bias related to the investigator, the sample, the method, or the instrument may not be completely avoidable in every case, it is important to know the possible sources of bias and how bias could affect the evidence. Moreover, scientists need to be attentive to possible bias in their own work as well as that of other scientists.

Objectivity may not always be attained. However, one precaution that may be taken to guard against undetected bias is to have many different investigators or groups of investigators working on a project. By different, it is meant that the groups are made up of various nationalities, ethnic origins, ages, and political convictions and composed of both males and females. It is also important to note one's aspirations, and to make sure to be truthful to the data, even when grants, promotions, and notoriety are at risk.

Skill 3.4   Recognize the basis for and application of safety practices and regulations in the study of science.

All science labs should contain the following items of **safety equipment**. Those marked with an asterisk are requirements by state laws.
* fire blanket which is visible and accessible
*Ground Fault Circuit Interrupters (GCFI) within two feet of water supplies
*signs designating room exits
*emergency shower capable of providing a continuous flow of water
*emergency eye wash station which can be activated by the foot or forearm
*eye protection for every student and a means of sanitizing equipment
*emergency exhaust fans providing ventilation to the outside of the building
*master cut-off switches for gas, electric and compressed air. Switches must have permanently attached handles. Cut-off switches must be clearly labeled.
*an ABC fire extinguisher
*storage cabinets for flammable materials
-chemical spill control kit
-fume hood with a motor which is spark proof
-protective laboratory aprons made of flame retardant material
-signs which will alert potential hazardous conditions
-containers for broken glassware, flammables, corrosives, and waste.
-containers should be labeled.

Students should wear safety goggles when performing dissections, heating, or while using acids and bases. Hair should always be tied back and objects should never be placed in the mouth. Food should not be consumed while in the laboratory. Hands should always be washed before and after laboratory experiments. In case of an accident, eye washes and showers should be used for eye contamination or a chemical spill that covers the student's body. Small chemical spills should only be contained and cleaned by the teacher. Kitty litter or a chemical spill kit should be used to clean spill. For large spills, the school administration and the local fire department should be notified. Biological spills should also be handled only by the teacher. Contamination with biological waste can be cleaned by using bleach when appropriate. Accidents and injuries should always be reported to the school administration and local health facilities. The severity of the accident or injury will determine the course of action to pursue.

It is the responsibility of the teacher to provide a safe environment for their students. Proper supervision greatly reduces the risk of injury and a teacher should never leave a class for any reason without providing alternate supervision. After an accident, two factors are considered: **foreseeability** and **negligence**. Foreseeability is the anticipation that an event may occur under certain circumstances. Negligence is the failure to exercise ordinary or reasonable care. Safety procedures should be a part of the science curriculum and a well-managed classroom is important to avoid potential lawsuits.

All laboratory solutions should be prepared as directed in the lab manual. Care should be taken to avoid contamination. All glassware should be rinsed thoroughly with distilled water before using and cleaned well after use. All solutions should be made with distilled water as tap water contains dissolved particles that may affect the results of an experiment. Unused solutions should be disposed of according to local disposal procedures.

The "Right to Know Law" covers science teachers who work with potentially hazardous chemicals. Briefly, the law states that employees must be informed of potentially toxic chemicals. An inventory must be made available if requested. The inventory must contain information about the hazards and properties of the chemicals. This inventory is to be checked against the "Substance List". Training must be provided on the safe handling and interpretation of the Material Safety Data Sheet.

The following chemicals are potential carcinogens and not allowed in school facilities: Acrylonitriel, Arsenic compounds, Asbestos, Bensidine, Benzene, Cadmium compounds, Chloroform, Chromium compounds, Ethylene oxide, Ortho-toluidine, Nickel powder, and Mercury.

Chemicals should not be stored on bench tops or heat sources. They should be stored in groups based on their reactivity with one another and in protective storage cabinets. All containers within the lab must be labeled. Suspect and known carcinogens must be labeled as such and segregated within trays to contain leaks and spills.

Chemical waste should be disposed of in properly labeled containers. Waste should be separated based on their reactivity with other chemicals.

Biological material should never be stored near food or water used for human consumption. All biological material should be appropriately labeled. All blood and body fluids should be put in a well-contained container with a secure lid to prevent leaking. All biological waste should be disposed of in biological hazardous waste bags.

Material safety data sheets are available for every chemical and biological substance. These are available directly from the company of acquisition or the internet. The manuals for equipment used in the lab should be read and understood before using them.

COMPETENCY 4.0 UNDERSTAND AND APPLY KNOWLEDGE OF THE INTERACTIONS AMONG SCIENCE, TECHNOLOGY, AND SOCIETY.

Skill 4.1 Recognize the historical and contemporary development of major scientific ideas and technological innovations.

Anton van Leeuwenhoek is known as the father of microscopy. In the 1650s, Leeuwenhoek began making tiny lenses, which gave magnifications up to 300x. He was the first to see and describe bacteria, yeast plants, and the microscopic life found in water. Over the years, light microscopes have advanced to produce greater clarity and magnification. The scanning electron microscope (SEM) was developed in the 1950s. Instead of light, a beam of electrons passes through the specimen. Scanning electron microscopes have a resolution about one thousand times greater than light microscopes. The disadvantage of the SEM is that the chemical and physical methods used to prepare the sample result in the death of the specimen.

In the late 1800s, Pasteur discovered the role of microorganisms in the cause of disease, pasteurization, and the rabies vaccine. Koch took this observations one step further by formulating that specific diseases were caused by specific pathogens. **Koch's postulates** are still used as guidelines in the field of microbiology. They state that the same pathogen must be found in every diseased person, the pathogen must be isolated and grown in culture, the disease is induced in experimental animals from the culture, and the same pathogen must be isolated from the experimental animal.

DNA structure was another key event in biological study. In the 1950s, James Watson and Francis Crick discovered the structure of a DNA molecule as that of a double helix. This structure made it possible to explain DNA's ability to replicate and to control the synthesis of proteins.

The use of animals in biological research has expedited many scientific discoveries. Animal research has allowed scientists to learn more about animal biological systems, including the circulatory and reproductive systems. One significant use of animals is for the testing of drugs, vaccines, and other products (such as perfumes and shampoos) before use or consumption by humans. Along with the pros of animal research, the cons are also very significant. The debate about the ethical treatment of animals has been ongoing since the introduction of animals in research. Many people believe the use of animals in research is cruel and unnecessary. Animal use is federally and locally regulated. The purpose of the Institutional Animal Care and Use Committee (IACUC) is to oversee and evaluate all aspects of an institution's animal care and use program.

Skill 4.2   Demonstrate an understanding of the ways that science and technology affect people's everyday lives, societal values and systems, the environment, and new knowledge.

Society as a whole impacts biological research. The pressure from the majority of society has led to these bans and restrictions on human cloning research. Human cloning has been restricted in the United States and many other countries. The U.S. legislature has banned the use of federal funds for the development of human cloning techniques. Some individual states have banned human cloning regardless of where the funds originate.

The demand for genetically modified crops by society and industry has steadily increased over the years. Genetic engineering in the agricultural field has led to improved crops for human use and consumption. Crops are genetically modified for increased growth and insect resistance because of the demand for larger and greater quantities of produce.

With advances in biotechnology come those in society who oppose it. Ethical questions come into play when discussing animal and human research. Does it need to be done? What are the effects on humans and animals? There are no right or wrong answers to these questions. There are governmental agencies in place to regulate the use of humans and animals for research.

Science and technology are often referred to as a "double-edged sword". Although advances in medicine have greatly improved the quality and length of life, certain moral and ethical controversies have arisen. Unforeseen environmental problems may result from technological advances. Advances in science have led to an improved economy through biotechnology as applied to agriculture, yet it has put our health care system at risk and has caused the cost of medical care to skyrocket. Society depends on science, yet is necessary that the public be scientifically literate and informed in order to allow potentially unethical procedures to occur. Especially vulnerable are the areas of genetic research and fertility. It is important for science teachers to stay abreast of current research and to involve students in critical thinking and ethics whenever possible.

Skill 4.3 **Analyze the processes of scientific and technological breakthroughs and their effects on other fields of study, careers, and job markets.**

Scientific and technological breakthroughs greatly influence other fields of study and the job market. All academic disciplines utilize computer and information technology to simplify research and information sharing. In addition, advances in science and technology influence the types of available jobs and the desired work Skills. For example, machines and computers continue to replace unskilled laborers and computer and technological literacy is now a requirement for many jobs and careers. Finally, science and technology continue to change the very nature of careers. Because of science and technology's great influence on all areas of the economy, and the continuing scientific and technological breakthroughs, careers are far less stable than in past eras. Workers can thus expect to change jobs and companies much more often than in the past.

Skill 4.4 **Analyze issues related to science and technology at the local, state, national, and global levels (e.g., environmental policies, genetic research).**

Local, state, national, and global governments and organizations must increasingly consider policy issues related to science and technology. For example, local and state governments must analyze the impact of proposed development and growth on the environment. Governments and communities must balance the demands of an expanding human population with the local ecology to ensure sustainable growth.

In addition, advances in science and technology create challenges and ethical dilemmas that national governments and global organizations must attempt to solve. Genetic research and manipulation, antibiotic resistance, stem cell research, and cloning are but a few of the issues facing national governments and global organizations.

In all cases, policy makers must analyze all sides of an issue and attempt to find a solution that protects society while limiting scientific inquiry as little as possible. For example, policy makers must weigh the potential benefits of stem cell research, genetic engineering, and cloning (e.g. medical treatments) against the ethical and scientific concerns surrounding these practices. Also, governments must tackle problems like antibiotic resistance, which can result from the indiscriminate use of medical technology (i.e. antibiotics), to prevent medical treatments from becoming obsolete.

Skill 4.5 **Evaluate the credibility of scientific claims made in various forums (e.g., the media, public debates, advertising).**

Because people often attempt to use scientific evidence in support of political or personal agendas, the ability to evaluate the credibility of scientific claims is a necessary Skill in today's society. In evaluating scientific claims made in the media, public debates, and advertising, one should follow several guidelines.

First, scientific, peer-reviewed journals are the most accepted source for information on scientific experiments and studies. One should carefully scrutinize any claim that does not reference peer-reviewed literature.

Second, the media and those with an agenda to advance (advertisers, debaters, etc.) often overemphasize the certainty and importance of experimental results. One should question any scientific claim that sounds fantastical or overly certain.

Finally, knowledge of experimental design and the scientific method is important in evaluating the credibility of studies. For example, one should look for the inclusion of control groups and the presence of data to support the given conclusions.

COMPETENCY 5.0 UNDERSTAND AND APPLY KNOWLEDGE OF THE MAJOR UNIFYING CONCEPTS OF ALL SCIENCES AND HOW THESE CONCEPTS RELATE TO OTHER DISCIPLINES.

Skill 5.1 Identify the major unifying concepts of the sciences (e.g., systems, order, and organization; constancy, change, and measurement) and their applications in real-life situations.

The following are the concepts and processes generally recognized as common to all scientific disciplines:

- Systems, order, and organization
- Evidence, models, and explanation
- Constancy, change, and measurement
- Evolution and equilibrium
- Form and function
-

Because the natural world is so complex, the study of science involves the **organization** of items into smaller groups based on interaction or interdependence. These groups are called **systems**. Examples of organization are the periodic table of elements and the five-kingdom classification scheme for living organisms. Examples of systems are the solar system, cardiovascular system, Newton's laws of force and motion, and the laws of conservation. **Order** refers to the behavior and measurability of organisms and events in nature. The arrangement of planets in the solar system and the life cycle of bacterial cells are examples of order.

Scientists use **evidence** and **models** to form **explanations** of natural events. Models are miniaturized representations of a larger event or system. Evidence is anything that furnishes proof.

Constancy and **change** describe the observable properties of natural organisms and events. Scientists use different systems of **measurement** to observe change and constancy. For example, the freezing and melting points of given substances and the speed of sound are constant under constant conditions. Growth, decay, and erosion are all examples of natural change.

Evolution is the process of change over a long period of time. While biological evolution is the most common example, one can also classify technological advancement, changes in the universe, and changes in the environment as evolution.

Equilibrium is the state of balance between opposing forces of change. Homeostasis and ecological balance are examples of equilibrium.

Form and **function** are properties of organisms and systems that are closely related. The function of an object usually dictates its form and the form of an object usually facilitates its function. For example, the form of the heart (e.g. muscle, valves) allows it to perform its function of circulating blood through the body.

Skill 5.2   Recognize connections within and among the traditional scientific disciplines.

Sciences are interconnected. The earth science discipline is closely related to biology, physics and chemistry.

Earth Science and Biology

Over the course of Earth's history, living things have been greatly affected by earth processes. Volcanic eruptions, plate tectonics, and climate change have affected whether living things have survived or how they have had to adapt in order to survive. At least four of the major mass extinctions have been caused by a climatic change triggered by some earth event, interconnecting geology, meteorology and biology. The working explanation for both the Cretaceous Tertiary extinction and the Permian-Triassic extinction include a large asteroid or meteorite impact. The majority of species prior to the impact were killed, and the remaining species had to adapt to the new environment in order to survive. Earth processes have also affected how humans live. The most fertile land for farming is at the base of volcanoes. We have developed technologies to irrigate farmland, built "safe" buildings in earthquake prone areas, and prevented low - lying areas from flooding. Currently, humans' impact on the environment and on the earth's temperature is being widely felt. All in all, biology has been greatly affected by earth processes.

Earth Science and Chemistry

Earth science and chemistry are tightly woven. In geology, the chemical composition of the rocks and the temperature and pressure at which crystals form are an obvious connection between chemistry and earth science. In addition, chemistry and oceanography are connected inherently. The salinity of earth's oceans is affected by the temperature at which water freezes, the density of water, and the solubility of certain chemical compounds. Chemistry and meteorology are connected through the chemical makeup of the atmosphere and the effects that human released chemicals have on the atmosphere (i.e. CFC's effect on the ozone layer and carbon dioxide's role in climate change.)

Earth Science and Physics

Earthquakes, plate tectonics, and meteorology are all related to physics. Fault production and earthquakes are caused by large scale plate tectonics forming small scale zones of weakness in the crust. The pressure builds up along fault lines due to increases in fault stress. At some point in time, the stress on the fault line will exceed the static frictional force of the fault line, and seismic waves will be released. The frictional force and the dynamics of the earth's motion during earthquakes are all related to physics.

Meteorology and physics are closely related. Changes in atmospheric pressure cause winds, updrafts and storms. These pressure changes are caused by changes in temperature. Warm, moist air rises because it is less dense than the air surrounding it. As air rises and cools, it condenses forming cloud systems. When meteorologists predict the weather, the physics of the interaction between volume, humidity, temperature and pressure is studied in detail.

Skill 5.3   Apply fundamental mathematical language, knowledge, and skills at the level of algebra and statistics in scientific contexts.

The knowledge and use of basic mathematical concepts and Skills is a necessary aspect of scientific study. Science depends on data and the manipulation of data requires knowledge of mathematics. Understanding of basic statistics, graphs and charts, and algebra are of particular importance. Scientists must be able to understand and apply the statistical concepts of mean, median, mode, and range to sets of scientific data. In addition, scientists must be able to represent data graphically and interpret graphs and tables. Finally, scientists often use basic algebra to solve scientific problems and design experiments. For example, the substitution of variables is a common strategy in experiment design. Also, the ability to determine the equation of a curve is valuable in data manipulation, experimentation, and prediction.

Modern science uses a number of disciplines to understand it better. Statistics is one of those subjects, which is absolutely essential for science.

Mean: Mean is the mathematical average of all the items. To calculate the mean, all the items must be added up and divided by the number of items. This is also called the arithmetic mean or more commonly as the "average".

Median: The median depends on whether the number of items is odd or even. If the number is odd, then the median is the value of the item in the middle. This is the value that denotes that the number of items having higher or equal value to that is same as the number of items having equal or lesser value than that. If the number o the items is even, the median is the average of the two items in the middle, such that the number of items having values higher or equal to it is same as the number o items having values equal or less than that.

Mode: Mode is the value o the item that occurs the most often, if there are not many items. Bimodal is a situation where there are two items with equal frequency.

Range: Range is the difference between the maximum and minimum values. The range is the difference between two extreme points on the distribution curve.

Scientists use mathematical tools and equations to model and solve scientific problems. Solving scientific problems often involves the use of quadratic, trigonometric, exponential, and logarithmic functions.

Quadratic equations take the standard form $ax^2 + bx + c = 0$. The most appropriate method of solving quadratic equations in scientific problems is the use of the quadratic formula. The quadratic formula produces the solutions of a standard form quadratic equation.

$$x = \frac{-b \pm \sqrt{b^2 - 4ac}}{2a} \quad \text{\{Quadratic Formula\}}$$

One common application of quadratic equations is the description of biochemical reaction equilibriums. Consider the following problem.

Example 1

80.0 g of ethanoic acid (MW = 60g) reacts with 85.0 g of ethanol (MW = 46g) until equilibrium. The equilibrium constant is 4.00. Determine the amounts of ethyl acetate and water produced at equilibrium.

$$CH_3COOH + CH_3CH_2OH = CH_3CO_2C_2H_5 + H_2O$$

The equilibrium constant, K, describes equilibrium of the reaction, relating the concentrations of products to reactants.

$$K = \frac{[CH_3CO_2C_2H_5][H_2O]}{[CH_3CO_2H][CH_3CH_2OH]} = 4.00$$

The equilibrium values of reactants and products are listed in the following table.

	CH_3COOH	CH_3CH_2OH	$CH_3CO_2C_2H_5$	H_2O
Initial	80/60 = 1.33 mol	85/46 = 1.85 mol	0	0
Equilibrium	1.33 − x	1.85 − x	x	x

Thus, $K = \dfrac{[x][x]}{[1.33-x][1.85-x]} = \dfrac{x^2}{2.46 - 3.18x + x^2} = 4.00$.

Rearrange the equation to produce a standard form quadratic equation.

$$\frac{x^2}{2.46 - 3.18x + x^2} = 4.00$$

$$x^2 = 4.00(2.46 - 3.18x + x^2) = 9.84 - 12.72x + 4x^2$$

$$0 = 3x^2 - 12.72x + 9.84$$

Use the quadratic formula to solve for x.

$$x = \frac{-(-12.72) \pm \sqrt{(-12.72)^2 - 4(3)(9.84)}}{2(3)} = 3.22 \text{ or } 1.02$$

3.22 is not an appropriate answer, because we started with only 3.18 moles of reactants. Thus, the amount of each product produced at equilibrium is 1.02 moles.

Scientists use trigonometric functions to define angles and lengths. For example, field biologists can use trigonometric functions to estimate distances and directions. The basic trigonometric functions are sine, cosine, and tangent. Consider the following triangle describing these relationships.

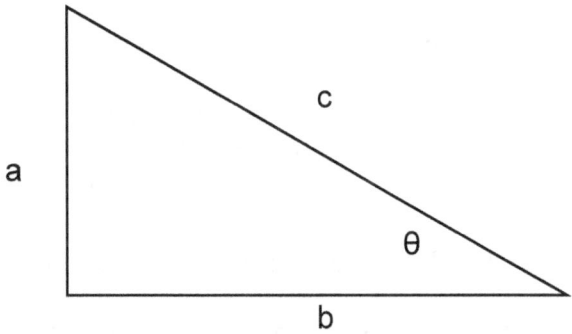

$$\sin \theta = \frac{a}{c}, \quad \cos \theta = \frac{b}{c}, \quad \tan \theta = \frac{a}{b}$$

Exponential functions are useful in modeling many scientific phenomena. For example, scientists use exponential functions to describe bacterial growth and radioactive decay. The general form of exponential equations is $f(x) = Ca^x$ (*C* is a constant). Consider the following problem involving bacterial growth.

Example 2

Determine the number of bacteria present in a culture inoculated with a single bacterium after 24 hours if the bacterial population doubles every 2 hours. Use $N(t) = N_0 e^{kt}$ as a model of bacterial growth where N(t) is the size of the population at time t, N_0 is the initial population size, and k is the growth constant.

We must first determine the growth constant, k. At t = 2, the size of the population doubles from 1 to 2. Thus, we substitute and solve for k.

$$2 = 1(e^{2k})$$

$$\ln 2 = \ln e^{2k} \quad \text{Take the natural log of each side.}$$

$$\ln 2 = 2k(\ln e) = 2k \quad \ln e = 1$$

$$k = \frac{\ln 2}{2} \quad \text{Solve for k.}$$

The population size at t = 24 is

$$N(24) = e^{(\frac{\ln 2}{2})24} = e^{12 \ln 2} = 4096.$$

Finally, logarithmic functions have many applications to science and biology. One simple example of a logarithmic application is the pH scale. Scientists define pH as follows.

pH = - $\log_{10}$ [H+], where [H+] is the concentration of hydrogen ions

Thus, we can determine the pH of a solution with a [H+] value of 0.0005 mol/L by using the logarithmic formula.

pH = - $\log_{10}$ [0.0005] = 3.3

Skill 5.4 Recognize the fundamental relationships among the natural sciences and the social sciences.

The fundamental relationship between the natural and social sciences is the use of the scientific method and the rigorous standards of proof that both disciplines require. This emphasis on organization and evidence separates the sciences from the arts and humanities. Natural science, particularly biology, is closely related to social science, the study of human behavior. Biological and environmental factors often dictate human behavior and accurate assessment of behavior requires a sound understanding of biological factors.

SUBAREA II.   LIFE SCIENCE

COMPETENCY 6.0   UNDERSTAND AND APPLY KNOWLEDGE OF CELL STRUCTURE AND FUNCTION.

Skill 6.1   **Compare and contrast the structures of viruses and prokaryotic and eukaryotic cells.**

The cell is the basic unit of all living things. There are three types of cells. They are prokaryotes, eukaryotes, and archaea. Archaea have some similarities with prokaryotes, but are as distantly related to prokaryotes as prokaryotes are to eukaryotes.

PROKARYOTES

Prokaryotes consist only of bacteria and cyanobacteria (formerly known as blue-green algae). The classification of prokaryotes is in the diagram below.

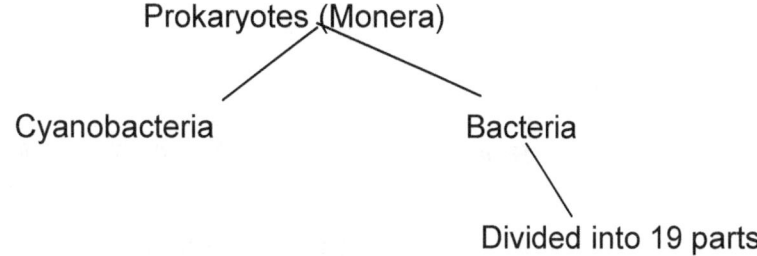

These cells have no defined nucleus or nuclear membrane. The DNA, RNA, and ribosomes float freely within the cell. The cytoplasm has a single chromosome condensed to form a **nucleoid**. Prokaryotes have a thick cell wall made up of amino sugars (glycoproteins). This is for protection, to give the cell shape, and to keep the cell from bursting. It is the **cell wall** of bacteria that is targeted by the antibiotic penicillin. Penicillin works by disrupting the cell wall, thus killing the cell.

The cell wall surrounds the **cell membrane** (plasma membrane). The cell membrane consists of a lipid bilayer that controls the passage of molecules in and out of the cell. Some prokaryotes have a capsule made of polysaccharides that surrounds the cell wall for extra protection from higher organisms.

Many bacterial cells have appendages used for movement called **flagella**. Some cells also have **pili**, which are a protein strand used for attachment of the bacteria. Pili may also be used for sexual conjugation (where the DNA from one bacterial cell is transferred to another bacterial cell).

Prokaryotes are the most numerous and widespread organisms on earth. Bacteria were most likely the first cells and date back in the fossil record to 3.5 billion years ago. Their ability to adapt to the environment allows them to thrive in a wide variety of habitats.

EUKARYOTES

Eukaryotic cells are found in protists, fungi, plants, and animals. Most eukaryotic cells are larger than prokaryotic cells. They contain many organelles, which are membrane bound areas for specific functions. Their cytoplasm contains a cytoskeleton which provides a protein framework for the cell. The cytoplasm also supports the organelles and contains the ions and molecules necessary for cell function. The cytoplasm is contained by the plasma membrane. The plasma membrane allows molecules to pass in and out of the cell. The membrane can bud inward to engulf outside material in a process called endocytosis. Exocytosis is a secretory mechanism, the reverse of endocytosis. The most significant differentiation between prokaryotes and eukaryotes is that eukaryotes have a **nucleus**.

VIRUSES

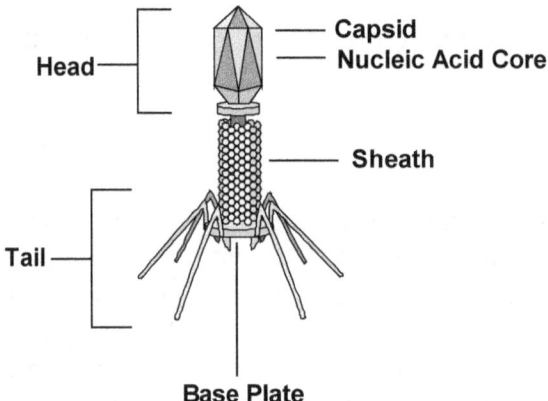

Bacteriophage

All viruses have a head or protein capsid that contains genetic material. This material is encoded in the nucleic acid and can be DNA, RNA, or even a limited number of enzymes. Some viruses also have a protein tail region. The tail aids in binding to the surface of the host cell and penetrating the surface of the host in order to introduce the virus's genetic material.

Other examples of viruses and their structures:

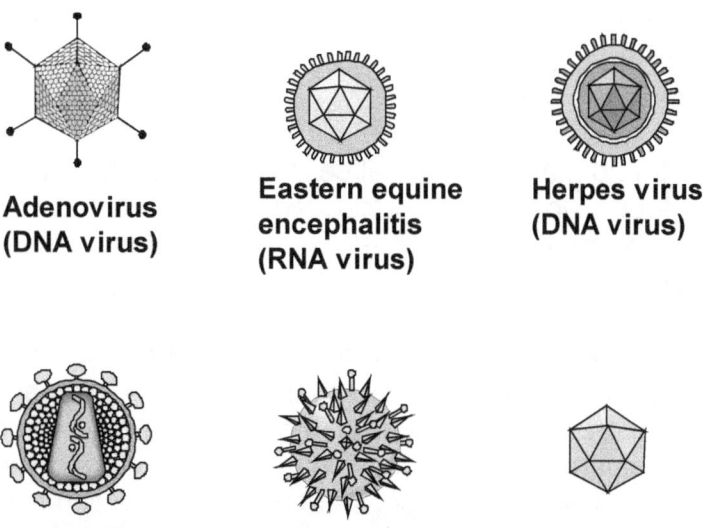

Adenovirus (DNA virus)

Eastern equine encephalitis (RNA virus)

Herpes virus (DNA virus)

HIV retrovirus (RNA virus)

Influenza virus (RNA virus)

Rotavirus (RNA virus)

Skill 6.2 Identify the structures and functions of cellular organelles.

The **nucleus** is the brain of the cell that contains all of the cell's genetic information. The chromosomes consist of chromatin, which is a complex of DNA and proteins. The chromosomes are tightly coiled to conserve space while providing a large surface area. The nucleus is the site of transcription of the DNA into RNA. The **nucleolus** is where ribosomes are made. There is at least one of these dark-staining bodies inside the nucleus of most eukaryotes. The nuclear envelope is two membranes separated by a narrow space. The envelope contains many pores that let RNA out of the nucleus.

Ribosomes are the site for protein synthesis. Ribosomes may be free floating in the cytoplasm or attached to the endoplasmic reticulum. There may be up to a half a million ribosomes in a cell, depending on how much protein is made by the cell.

The **endoplasmic reticulum** (ER) is folded and provides a large surface area. It is the "roadway" of the cell and allows for transport of materials through and out of the cell. There are two types of ER. Smooth endoplasmic reticulum contains no ribosomes on their surface. This is the site of lipid synthesis. Rough endoplasmic reticulum has ribosomes on their surface. They aid in the synthesis of proteins that are membrane bound or destined for secretion.

Many of the products made in the ER proceed on to the Golgi apparatus. The **Golgi apparatus** functions to sort, modify, and package molecules that are made in the other parts of the cell (like the ER). These molecules are either sent out of the cell or to other organelles within the cell. The Golgi apparatus is a stacked structure to increase the surface area.

Lysosomes are found mainly in animal cells. These contain digestive enzymes that break down food, substances not needed, viruses, damaged cell components and eventually the cell itself. It is believed that lysomomes are responsible for the aging process.

Mitochondria are large organelles that are the site of cellular respiration, where ATP is made to supply energy to the cell. Muscle cells have many mitochondria because they use a great deal of energy. Mitochondria have their own DNA, RNA, and ribosomes and are capable of reproducing by binary fission if there is a greater demand for additional energy. Mitochondria have two membranes: a smooth outer membrane and a folded inner membrane. The folds inside the mitochondria are called cristae. They provide a large surface area for cellular respiration to occur.

Plastids are found only in photosynthetic organisms. They are similar to the mitochondira due to the double membrane structure. They also have their own DNA, RNA, and ribosomes and can reproduce if the need for the increased capture of sunlight becomes necessary. There are several types of plastids. **Chloroplasts** are the sight of photosynthesis. The stroma is the chloroplast's inner membrane space. The stoma encloses sacs called thylakoids that contain the photosynthetic pigment chlorophyll. The chlorophyll traps sunlight inside the thylakoid to generate ATP that is used in the stroma to produce carbohydrates and other products. The **chromoplasts** make and store yellow and orange pigments. They provide color to leaves, flowers, and fruits. The **amyloplasts** store starch and are used as a food reserve. They are abundant in roots like potatoes.

The Endosymbiotic Theory states that mitochondria and chloroplasts were once free living and possibly evolved from prokaryotic cells. At some point in our evolutionary history, they entered the eukaryotic cell and maintained a symbiotic relationship with the cell, with both the cell and organelle benefiting from the relationship. The fact that they both have their own DNA, RNA, ribosomes, and are capable of reproduction helps to confirm this theory.

Found in plant cells only, the **cell wall** is composed of cellulose and fibers. It is thick enough for support and protection, yet porous enough to allow water and dissolved substances to enter. **Vacuoles** are found mostly in plant cells. They hold stored food and pigments. Their large size allows them to fill with water in order to provide turgor pressure. Lack of turgor pressure causes a plant to wilt.

The **cytoskeleton**, found in both animal and plant cells, is composed of protein filaments attached to the plasma membrane and organelles. They provide a framework for the cell and aid in cell movement. They constantly change shape and move about. Three types of fibers make up the cytoskeleton:

1. **Microtubules** – the largest of the three, they make up cilia and flagella for locomotion. Some examples are sperm cells, cilia that line the fallopian tubes and tracheal cilia. Centrioles are also composed of microtubules. They aid in cell division to form the spindle fibers that pull the cell apart into two new cells. Centrioles are not found in the cells of higher plants.

2. **Intermediate filaments** – intermediate in size, they are smaller than microtubules but larger than microfilaments. They help the cell to keep its shape.

3. **Microfilaments** – smallest of the three, they are made of actin and small amounts of myosin (like in muscle tissue). They function in cell movement like cytoplasmic streaming, endocytosis, and ameboid movement. This structure pinches the two cells apart after cell division, forming two new cells.

The following is a diagram of a generalized animal cell.

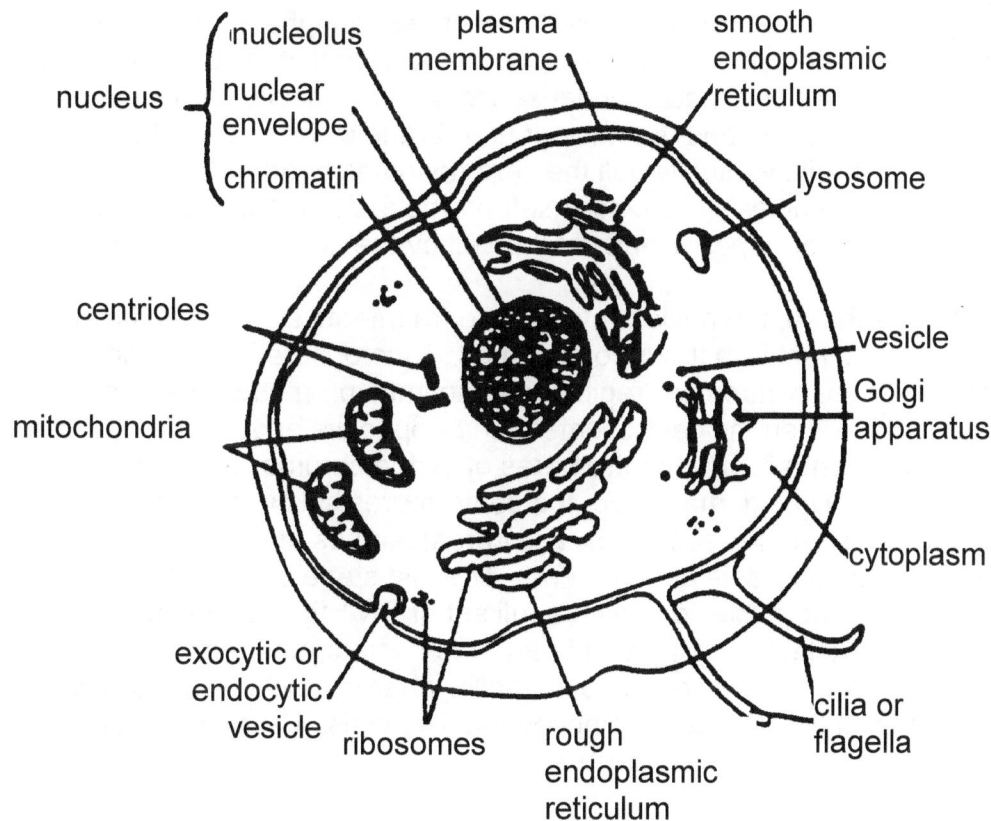

ARCHAEA

There are three kinds of organisms with archaea cells: **methanogens** are obligate anaerobes that produce methane, **halobacteria** can live only in concentrated brine solutions, and **thermoacidophiles** can only live in acidic hot springs.

Skill 6.3 **Describe the processes of the cell cycle.**

The purpose of cell division is to provide growth and repair in body (somatic) cells and to replenish or create sex cells for reproduction. There are two forms of cell division. **Mitosis** is the division of somatic cells and **meiosis** is the division of sex cells (eggs and sperm).

Mitosis is divided into two parts: the **mitotic (M) phase** and **interphase**. In the mitotic phase, mitosis and cytokinesis divide the nucleus and cytoplasm, respectively. This phase is the shortest phase of the cell cycle. Interphase is the stage where the cell grows and copies the chromosomes in preparation for the mitotic phase. Interphase occurs in three stages of growth: **G1** (growth) period is when the cell is growing and metabolizing, the **S** period (synthesis) is where new DNA is being made and the **G2** phase (growth) is where new proteins and organelles are being made to prepare for cell division.

The mitotic phase is a continuum of change, although it is described as occurring in five stages: prophase, prometaphase, metaphase, anaphase, and telophase. During **prophase**, the cell proceeds through the following steps continuously, with no stopping. The chromatin condenses to become visible chromosomes. The nucleolus disappears and the nuclear membrane breaks apart. Mitotic spindles form that will eventually pull the chromosomes apart. They are composed of microtubules. The cytoskeleton breaks down and the spindles are pushed to the poles or opposite ends of the cell by the action of centrioles.

During **prometaphase**, the nuclear membrane fragments and allows the spindle microtubules to interact with the chromosomes. Kinetochore fibers attach to the chromosomes at the centromere region. (Sometimes prometaphase is grouped with metaphase). When the centrosomes are at opposite ends of the cell, the division is in **metaphase**. The centromeres of all the chromosomes are aligned with one another. During **anaphase**, the centromeres split in half and homologous chromosomes separate. The chromosomes are pulled to the poles of the cell, with identical sets at either end. The last stage of mitosis is **telophase**. Here, two nuclei form with a full set of DNA that is identical to the parent cell. The nucleoli become visible and the nuclear membrane reassembles. A cell plate is seen in plant cells, whereas a cleavage furrow is formed in animal cells. The cell is pinched into two cells. Cytokinesis, or division of the cytoplasm and organelles, occurs.

Below is a diagram of mitosis.

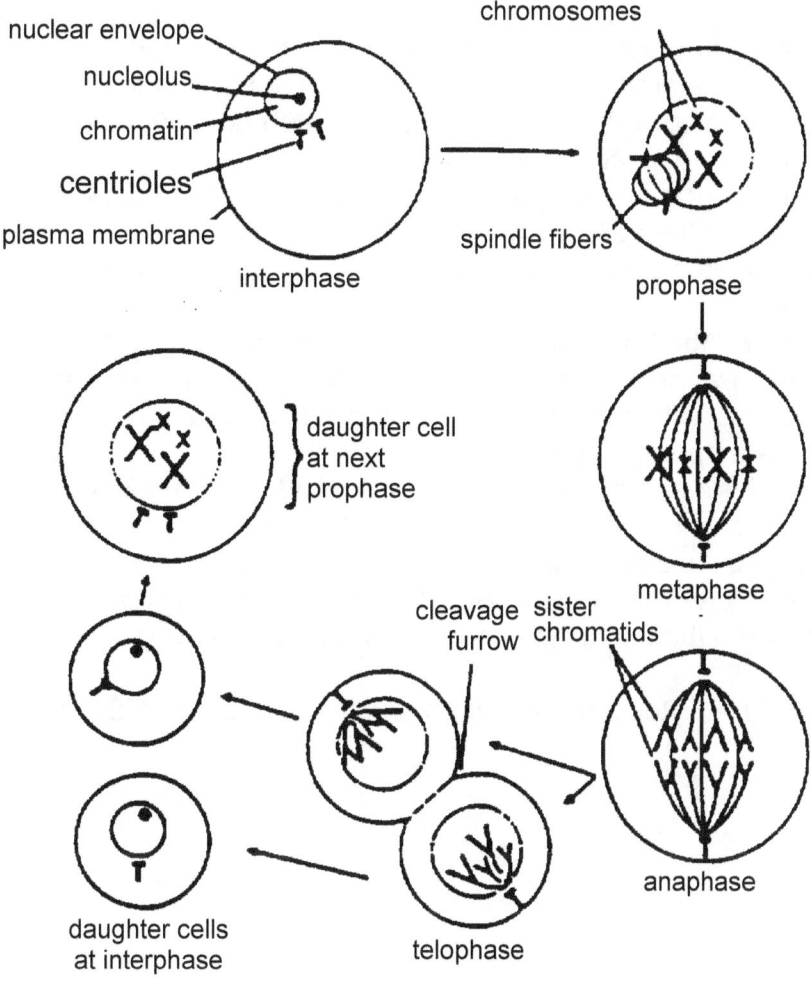

Mitosis in an Animal Cell

Meiosis is similar to mitosis, but there are two consecutive cell divisions, meiosis I and meiosis II in order to reduce the chromosome number by one half. This way, when the sperm and egg join during fertilization, the haploid number is reached.

Similar to mitosis, meiosis is preceded by an interphase during which the chromosome replicates. The steps of meiosis are as follows:

1. **Prophase I** – the replicated chromosomes condense and pair with homologues in a process called synapsis. This forms a tetrad. Crossing over, the exchange of genetic material between homologues to further increase diversity, occurs during prophase I.
2. **Metaphase I** – the homologous pairs attach to spindle fibers after lining up in the middle of the cell.
3. **Anaphase I** – the sister chromatids remain joined and move to the poles of the cell.
4. **Telophase I** – the homologous chromosome pairs continue to separate. Each pole now has a haploid chromosome set. Telophase I occurs simultaneously with cytokinesis. In animal cells, cleavage furrows form and cell plate appear in plant cells.
5. **Prophase II** – a spindle apparatus forms and the chromosomes condense.
6. **Metaphase II** – sister chromatids line up in center of cell. The centromeres divide and the sister chromatids begin to separate.
7. **Anaphase II** – the separated chromosomes move to opposite ends of the cell.
8. **Telophase II** – cytokinesis occurs, resulting in four haploid daughter cells.

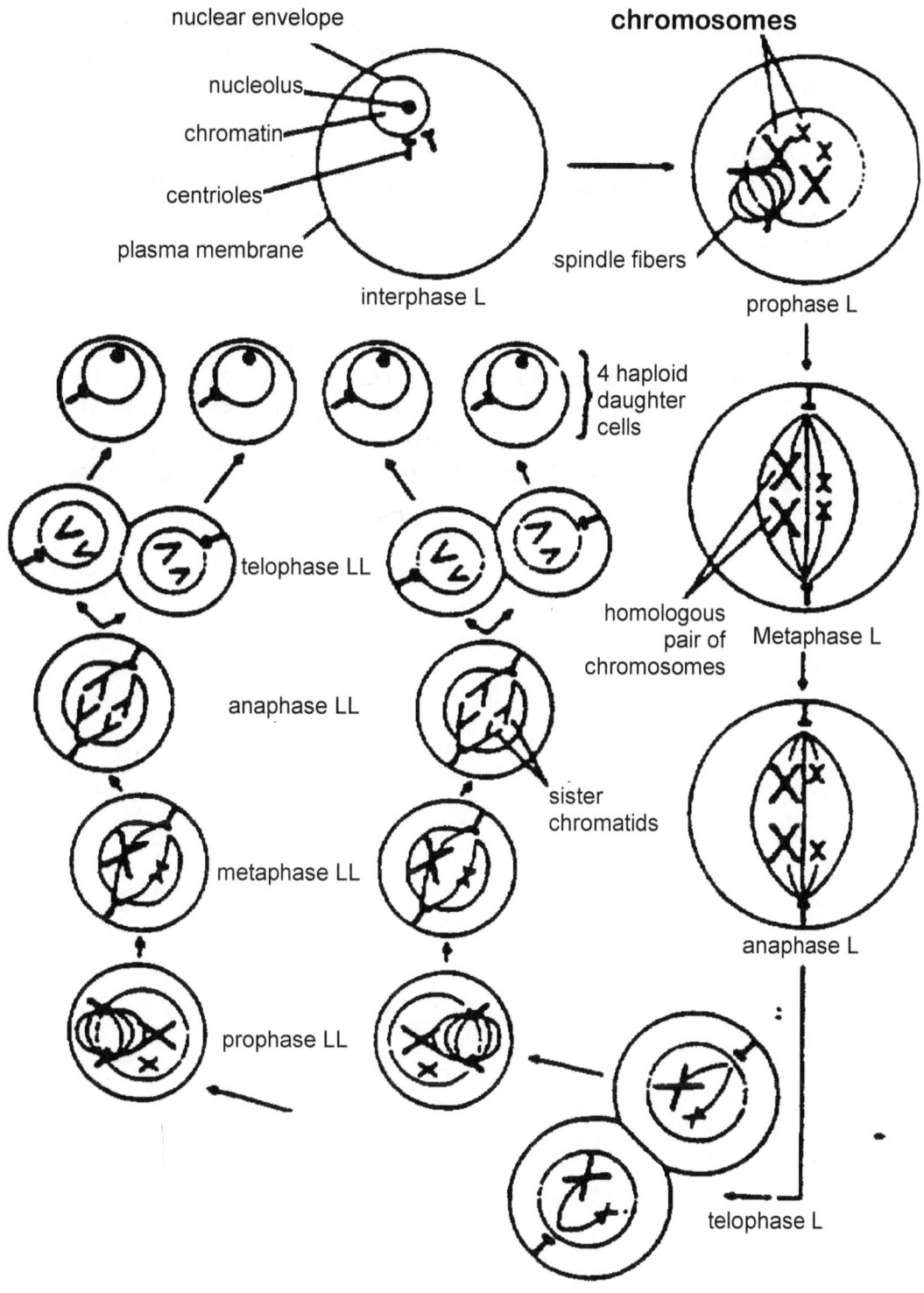

Skill 6.4 **Explain the functions and applications of the instruments and technologies used to study the life sciences at the molecular and cellular level.**

Gel electrophoresis is a method for analyzing DNA. Electrophoresis separates DNA or protein by size or electrical charge. The DNA runs towards the positive charge as it separates the DNA fragments by size. The gel is treated with a DNA-binding dye that fluoresces under ultraviolet light. A picture of the gel can be taken and used for analysis.

One of the most widely used genetic engineering techniques is **polymerase chain reaction (PCR)**. PCR is a technique in which a piece of DNA can be amplified into billions of copies within a few hours. This process requires primer to specify the segment to be copied, and an enzyme (usually taq polymerase) to amplify the DNA. PCR has allowed scientists to perform several procedures on the smallest amount of DNA.

COMPETENCY 7.0 UNDERSTAND AND APPLY KNOWLEDGE OF THE PRINCIPLES OF HEREDITY AND BIOLOGICAL EVOLUTION.

Skill 7.1 Recognize the nature and function of the gene, with emphasis on the molecular basis of inheritance and gene expression.

Gregor Mendel is recognized as the father of genetics. His work in the late 1800s is the basis of our knowledge of genetics. Although unaware of the presence of DNA or genes, Mendel realized there were factors (now known as **genes**) that were transferred from parents to their offspring. Mendel worked with pea plants and fertilized the plants himself, keeping track of subsequent generations which led to the Mendelian laws of genetics. Mendel found that two "factors" governed each trait, one from each parent. Traits or characteristics came in several forms, known as **alleles**. For example, the trait of flower color had white alleles (*pp*) and purple alleles (*PP*). Mendel formed two laws: the law of segregation and the law of independent assortment.

In bacterial cells, the *lac* operon is a good example of the control of gene expression. The *lac* operon contains the genes that encode for the enzymes used to convert lactose into fuel (glucose and galactose). The *lac* operon contains three genes, *lac Z*, *lac Y*, and *lac A*. *Lac Z* encodes an enzyme for the conversion of lactose into glucose and galactose. *Lac Y* encodes for an enzyme that causes lactose to enter the cell. *Lac A* encodes for an enzyme that acetylates lactose.

The *lac* operon also contains a promoter and an operator that is the "off and on" switch for the operon. A protein called the repressor switches the operon off when it binds to the operator. When lactose is absent, the repressor is active and the operon is turned off. The operon is turned on again when allolactose (formed from lactose) inactivates the repressor by binding to it.

Skill 7.2   Analyze the transmission of genetic information (e.g., Punnett squares, sex-linked traits, pedigree analysis).

The **law of segregation** states that only one of the two possible alleles from each parent is passed on to the offspring. If the two alleles differ, then one is fully expressed in the organism's appearance (the dominant allele) and the other has no noticeable effect on appearance (the recessive allele). The two alleles for each trait segregate into different gametes. A Punnet square can be used to show the law of segregation. In a Punnet square, one parent's genes are put at the top of the box and the other parent's on the side. Genes combine in the squares just like numbers are added in addition tables. This Punnet square shows the result of the cross of two F_1 hybrids.

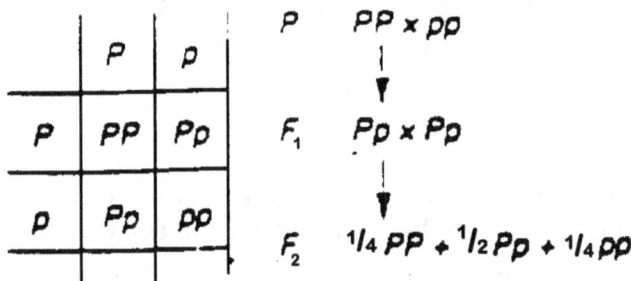

This cross results in a 1:2:1 ratio of F_2 offspring. Here, the *P* is the dominant allele and the *p* is the recessive allele. The F_1 cross produces three offspring with the dominant allele expressed (two *PP* and *Pp*) and one offspring with the recessive allele expressed (*pp*). Some other important terms to know:

> **Homozygous** – having a pair of identical alleles. For example, *PP* and *pp* are homozygous pairs.
> **Heterozygous** – having two different alleles. For example, *Pp* is a heterozygous pair.
> **Phenotype** – the organism's physical appearance.
> **Genotype** – the organism's genetic makeup. For example, *PP* and *Pp* have the same phenotype (purple in color), but different genotypes.

The **law of independent assortment** states that alleles sort independently of each other. The law of segregation applies for a monohybrid crosses (only one character, in this case flower color, is experimented with). In a dihybrid cross, two characters are being explored. Two of the seven characters Mendel studied were seed shape and color. Yellow is the dominant seed color (*Y*) and green is the recessive color (*y*). The dominant seed shape is round (*R*) and the recessive shape is wrinkled (*r*). A cross between a plant with yellow round seeds (*YYRR*) and a plant with green wrinkled seeds (*yyrr*) produces an F_1 generation with the genotype *YyRr*. The production of F_2 offspring results in a 9:3:3:1 phenotypic ratio.

F₂

	YR	Yr	yR	yr
YR	YYRR	YYRr	YyRR	YyRr
Yr	YYRr	YYrr	YyRr	Yyrr
yR	YyRR	YyRr	yyRR	yyRr
yr	YyRr	Yyrr	yyRr	yyrr

P   YYRR × yyrr

F₁   YyRr

F₂   YYRR – 1
       YYRr – 2
       YyRR – 2
       YyRr – 4   } 9 yellow round

       yyRR – 1
       yyRr – 2   } 3 green round

       YYrr – 1
       Yyrr – 2   } 3 yellow wrinkled

       yyrr – 1   } 1 green wrinkled

Based on Mendelian genetics, the more complex hereditary pattern of **dominance** was discovered. In Mendel's law of segregation, the F_1 generation have either purple or white flowers. This is an example of **complete dominance**. **Incomplete dominance** is when the F_1 generation results in an appearance somewhere between the two parents. For example, red flowers are crossed with white flowers, resulting in an F_1 generation with pink flowers. The red and white traits are still carried by the F_1 generation, resulting in an F_2 generation with a phenotypic ration of 1:2:1. In **codominance,** the genes may form new phenotypes. The ABO blood grouping is an example of codominance. A and B are of equal strength and O is recessive. Therefore, type A blood may have the genotypes of AA or AO, type B blood may have the genotypes of BB or BO, type AB blood has the genotype A and B, and type O blood has two recessive O genes.

A **family pedigree** is a collection of a family's history for a particular trait. As you work your way through the pedigree of interest, the Mendelian inheritance theories are applied. In tracing a trait, the generations are mapped in a pedigree chart, similar to a family tree but with the alleles present. In a case where both parent have a particular trait and one of two children also express this trait, then the trait is due to a dominant allele. In contrast, if both parents do not express a trait and one of their children do, that trait is due to a recessive allele.

Sex linked traits - the Y chromosome found only in males (XY) carries very little genetic information, whereas the X chromosome found in females (XX) carries very important information. Since men have no second X chromosome to cover up a recessive gene, the recessive trait is expressed more often in men. Women need the recessive gene on both X chromosomes to show the trait. Examples of sex linked traits include hemophilia and color-blindness.

Skill 7.3 Analyze the processes of change at the microscopic and Macroscopic levels.

In order to fully understand heredity and biological evolution, students need to comprehend the material on both a smaller (microscopic) and a larger (macroscopic) scale. For example, smaller items would include molecules, DNA, and genes and larger items might include organisms and the biosphere.

The teaching of molecular biology is important to the understanding of the chemical basis of life. Students tend to associate molecules with physical science, not realizing that living systems are made of molecules as well as cells. At the macroscopic level, students learn species as a basis for classifying organisms; in order to understand species at the microscopic level, they need to comprehend the genetic basis of the species.

Microscopic Level:

Atoms, molecules, chemical processes and reactions, bacteria, viruses, protists, cells, tissues, chromosomes, genes, meiosis, mitosis, mutations, comparative embryology, molecular evolution.

Macroscopic Level:

Comparative anatomy, natural selection, convergent evolution, divergent evolution, taxonomy, organs, organisms, body systems, animal and plant structure and function, animal behavior, populations, communities, food chains and webs, biomes.

Skill 7.4 **Identify scientific evidence from various sources, such as the fossil record, comparative anatomy, and biochemical similarities, to demonstrate knowledge of theories about processes of biological evolution.**

The hypothesis that life developed on Earth from nonliving materials is the most widely accepted theory on the origin of life. The transformation from nonliving materials to life had four stages. The first stage was the nonliving (abiotic) synthesis of small monomers such as amino acids and nucleotides. In the second stage, these monomers combine to form polymers, such as proteins and nucleic acids. The third stage is the accumulation of these polymers into droplets called protobionts. The last stage is the origin of heredity, with RNA as the first genetic material.

The first stage of this theory was hypothesized in the 1920s. A. I. Oparin and J. B. S. Haldane were the first to theorize that the primitive atmosphere was a reducing atmosphere with no oxygen present. The gases were rich in hydrogen, methane, water and ammonia.. In the 1950s, Stanley Miller proved Oparin's theory in the laboratory by combining the above gases. When given an electrical spark, he was able to synthesize simple amino acids. It is commonly accepted that amino acids appeared before DNA. Other laboratory experiments have supported the other stages in the origin of life theory could have happened.

Other scientists believe simpler hereditary systems originated before nucleic acids. In 1991, Julius Rebek was able to synthesize a simple organic molecule that replicates itself. According to his theory, this simple molecule may be the precursor of RNA.

Prokaryotes are the simplest life form. Their small genome size limits the number of genes that control metabolic activities. Over time, some prokaryotic groups became multicellular organisms for this reason. Prokaryotes then evolved to form complex bacterial communities where species benefit from one another.

The **endosymbiotic theory** of the origin of eukaryotes states that eukaryotes arose from symbiotic groups of prokaryotic cells. According to this theory, smaller prokaryotes lived within larger prokaryotic cells, eventually evolving into chloroplasts and mitochondria. Chloroplasts are the descendant of photosynthetic prokaryotes and mitochondria are likely to be the descendants of bacteria that were aerobic heterotrophs. Serial endosymbiosis is a sequence of endosymbiotic events. Serial endosymbiosis may also play a role in the progression of life forms to become eukaryotes.

Fossils are the key to understanding biological history. They are the preserved remnants left by an organism that lived in the past. Scientists have established the geological time scale to determine the age of a fossil. The geological time scale is broken down into four eras: the Precambrian, Paleozoic, Mesozoic, and Cenozoic. The eras are further broken down into periods that represent a distinct age in the history of Earth and its life. Scientists use rock layers called strata to date fossils. The older layers of rock are at the bottom. This allows scientists to correlate the rock layers with the era they date back to. Radiometric dating is a more precise method of dating fossils. Rocks and fossils contain isotopes of elements accumulated over time. The isotope's half-life is used to date older fossils by determining the amount of isotope remaining and comparing it to the half-life.

Dating fossils is helpful to construct and evolutionary tree. Scientists can arrange the succession of animals based on their fossil record. The fossils of an animal's ancestors can be dated and placed on its evolutionary tree. For example, the branched evolution of horses shows the progression of the modern horse's ancestors to be larger, to have a reduced number of toes, and have teeth modified for grazing.

Comparative anatomical studies reveal that some structural features are basically similar – e.g., flowers generally have sepals, petals, stigma, style and ovary but the size, color, number of petals, sepals etc., may differ from species to species.

The degree of resemblance between two organisms indicates how closely they are related in evolution.

- Groups with little in common are supposed to have diverged from a common ancestor much earlier in geological history than groups which have more in common

- To decide how closely two organisms are, anatomists look for the structures which may serve different purpose in the adult, but are basically similar (homologous)

- In cases where similar structures serve different functions in adults, it is important to trace their origin and embryonic development

When a group of organisms share a homologous structure, which is specialized, to perform a variety of functions in order to adapt to different environmental conditions are called adaptive radiation. The gradual spreading of organisms with adaptive radiation is known as divergent evolution.

Examples of divergent evolution are pentadactyl limb and insect mouthparts.

Under similar environmental conditions, fundamentally different structures in different groups of organisms may undergo modifications to serve similar functions. This is called convergent evolution. The structures, which have no close phylogenetic links but showing adaptation to perform the same functions, are called analogous.

Examples are – wings of bats, bird and insects, jointed legs of insects and vertebrates, eyes of vertebrates and cephalopods.

Vestigial organs: Organs that are smaller and simpler in structure than corresponding parts in the ancestral species are called vestigial organs. They are usually degenerated or underdeveloped. These were functional in ancestral species but no have become non functional, e.g., vestigial hind limbs of whales, vestigial leaves of some xerophytes, vestigial wings of flightless birds like ostriches, etc.

COMPETENCY 8.0 UNDERSTAND AND APPLY KNOWLEDGE OF THE CHARACTERISTICS AND LIFE FUNCTIONS OF ORGANISMS.

Skill 8.1 Identify the levels of organization of various types of organisms and the structures and functions of cells, tissues, organs, and organ systems.

Life has defining properties. Some of the more important processes and properties associated with life are as follows:

*Order – an organism's complex organization.
*Reproduction – life only comes from life (biogenesis).
*Energy utilization – organisms use and make energy to do many kinds of work.
*Growth and development – DNA directed growth and development.
*Adaptation to the environment – occurs by homeostasis (ability to maintain a certain status), response to stimuli, and evolution.

Life is highly organized. The organization of living systems builds on levels from small to increasingly more large and complex. All aspects, whether it is a cell or an ecosystem, have the same requirements to sustain life. Life is organized from simple to complex in the following way:

Atoms ® molecules ® organelles ® cells ® tissues ® organs ® organ systems ® organism

Skill 8.2 Analyze the strategies and adaptations used by organisms to obtain the basic requirements of life.

Members of the five different kingdoms of the classification system of living organisms often differ in their basic life functions. Here we compare and analyze how members of the five kingdoms obtain nutrients, excrete waste, and reproduce.

Bacteria are prokaryotic, single-celled organisms that lack cell nuclei. The different types of bacteria obtain nutrients in a variety of ways. Most bacteria absorb nutrients from the environment through small channels in their cell walls and membranes (chemotrophs) while some perform photosynthesis (phototrophs). Chemoorganotrophs use organic compounds as energy sources while chemolithotrophs can use inorganic chemicals as energy sources. Depending on the type of metabolism and energy source, bacteria release a variety of waste products (e.g. alcohols, acids, carbon dioxide) to the environment through diffusion.

All bacteria reproduce through binary fission (asexual reproduction) producing two identical cells. Bacteria reproduce very rapidly, dividing or doubling every twenty minutes in optimal conditions. Asexual reproduction does not allow for genetic variation, but bacteria achieve genetic variety by absorbing DNA from ruptured cells and conjugating or swapping chromosomal or plasmid DNA with other cells.

Animals are multicellular, eukaryotic organisms. All animals obtain nutrients by eating food (ingestion). Different types of animals derive nutrients from eating plants, other animals, or both. Animal cells perform respiration that converts food molecules, mainly carbohydrates and fats, into energy. The excretory systems of animals, like animals themselves, vary in complexity. Simple invertebrates eliminate waste through a single tube, while complex vertebrates have a specialized system of organs that process and excrete waste.

Most animals, unlike bacteria, exist in two distinct sexes. Members of the female sex give birth or lay eggs. Some less developed animals can reproduce asexually. For example, flatworms can divide in two and some unfertilized insect eggs can develop into viable organisms. Most animals reproduce sexually through various mechanisms. For example, aquatic animals reproduce by external fertilization of eggs, while mammals reproduce by internal fertilization. More developed animals possess specialized reproductive systems and cycles that facilitate reproduction and promote genetic variation.

Plants, like animals, are multi-cellular, eukaryotic organisms. Plants obtain nutrients from the soil through their root systems and convert sunlight into energy through photosynthesis. Many plants store waste products in vacuoles or organs (e.g. leaves, bark) that are discarded. Some plants also excrete waste through their roots.

More than half of the plant species reproduce by producing seeds from which new plants grow. Depending on the type of plant, flowers or cones produce seeds. Other plants reproduce by spores, tubers, bulbs, buds, and grafts. The flowers of flowering plants contain the reproductive organs. Pollination is the joining of male and female gametes that is often facilitated by movement by wind or animals.

Fungi are eukaryotic, mostly multi-cellular organisms. All fungi are heterotrophs, obtaining nutrients from other organisms. More specifically, most fungi obtain nutrients by digesting and absorbing nutrients from dead organisms. Fungi secrete enzymes outside of their body to digest organic material and then absorb the nutrients through their cell walls.

Most fungi can reproduce asexually and sexually. Different types of fungi reproduce asexually by mitosis, budding, sporification, or fragmentation. Sexual reproduction of fungi is different from sexual reproduction of animals. The two mating types of fungi are plus and minus, not male and female. The fusion of hyphae, the specialized reproductive structure in fungi, between plus and minus types produces and scatters diverse spores.

Protists are eukaryotic, single-celled organisms. Most protists are heterotrophic, obtaining nutrients by ingesting small molecules and cells and digesting them in vacuoles. All protists reproduce asexually by either binary or multiple fission. Like bacteria, protists achieve genetic variation by exchange of DNA through conjugation.

Skill 8.3　　Analyze factors (e.g., physiological, behavioral) that influence homeostasis within an organism.

Animal behavior is responsible for courtship leading to mating, communication between species, territoriality, and aggression between animals and dominance within a group. Behaviors may include body posture, mating calls, display of feathers or fur, coloration or bearing of teeth and claws.

Innate behaviors are inborn or instinctual. An environmental stimulus such as the length of day or temperature results in a behavior. Hibernation among some animals is an innate behavior, as is a change in color known as camouflage.

Learned behavior is modified behavior due to past experience.

Skill 8.4　　Demonstrate an understanding of the human as a living organism with life functions comparable to those of other life forms.

Humans are living organisms that display the basic properties of life. Humans share functional characteristics with all living organisms, from simple bacteria to complex mammals. The basic functions of living organisms include reproduction, growth and development, metabolism, and homeostasis/response to the environment.

Reproduction – All living organisms reproduce their own kind. Life arises only from other life. Humans reproduce through sexual reproduction, requiring the interaction of a male and a female. Human sexual reproduction is nearly identical to reproduction in other mammals. In addition, while simpler organisms have different methods of reproduction, they all reproduce. For example, the major mechanism of bacterial reproduction is asexual binary fission in which the cell divides in half, producing two identical cells.

Growth and Development – Growth and development, as directed by DNA, produces an organism characteristic of its species. In humans and other higher-level mammals, growth and development is a very complex process. In humans, growth and development requires differentiation of cells into many different types to form the various organs, structures, and functional elements. While differentiation is unique to higher level organisms, all living organisms grow. For example, the simplest bacterial cell grows in size until it divides into two organisms. Human body cells undergo a similar process, growing in size until division is necessary.

Metabolism – Metabolism is the sum of all chemical reactions that occur in a living organism. Catabolism is the breaking down of complex molecules to release energy. Anabolism is the utilization of the energy from catabolism to build complex molecules. Cellular respiration, the basic mechanism of catabolism in humans, is common to many living organisms of varying levels of complexity.

Homeostasis/Response to the Environment – All living organisms respond and adapt to their environments. Homeostasis is the result of regulatory mechanisms that help maintain an organism's internal environment within tolerable limits. For example, in humans and mammals, constriction and dilation of blood vessels near the skin help maintain body temperature.

COMPETENCY 9.0 UNDERSTAND AND APPLY KNOWLEDGE OF HOW ORGANISMS INTERACT WITH EACH OTHER AND WITH THEIR ENVIRONMENT.

Skill 9.1 Identify living and nonliving components of the environment and how they interact with one another.

Succession is an orderly process of replacing a community that has been damaged or has begun where no life previously existed. Primary succession occurs where life never existed before, as in a flooded area or a new volcanic island. Secondary succession takes place in communities that were once flourishing but disturbed by some source, either man or nature, but not totally stripped. A climax community is a community that is established and flourishing.

Abiotic and biotic factors play a role in succession. **Biotic factors** are living things in an ecosystem: plants, animals, bacteria, fungi, etc. **Abiotic factors** are non-living aspects of an ecosystem: soil quality, rainfall, temperature, etc.

Abiotic factors affect succession by way of the species that colonize the area. Certain species will or will not survive depending on the weather, climate, or soil makeup. Biotic factors such as inhibition of one species due to another may occur. This may be due to some form of competition between the species.

Skill 9.2 Recognize the concepts of populations, communities, ecosystems, and ecoregions and the role of biodiversity in living systems.

Ecology is the study of organisms, where they live and their interactions with the environment. A **population** is a group of the same species in a specific area. Many factors can affect the population size and its growth rate. Population size can depend on the total amount of life a habitat can support. This is the carrying capacity of the environment. Once the habitat runs out of food, water, shelter, or space, the carrying capacity decreases, and then stabilizes. A **community** is a group of populations residing in the same area. Communities that are ecologically similar in regards to temperature, rainfall and the species that live there are called **biomes**. Specific biomes include:

> **Marine** - covers 75% of the earth. This biome is organized by the depth of the water. The intertidal zone is from the tide line to the edge of the water. The littoral zone is from the water's edge to the open sea. It includes coral reef habitats and is the most densely populated area of the marine biome. The open sea zone is divided into the epipelagic zone and the pelagic zone. The epipelagic zone receives more sunlight and has a larger number of species. The ocean floor is called the benthic zone and is populated with bottom feeders.

Tropical Rain Forest - temperature is constant (25 degrees C), rainfall exceeds 200 cm. per year. Located around the area of the equator, the rain forest has abundant, diverse species of plants and animals.

Savanna - temperatures range from 0-25 degrees C depending on the location. Rainfall is from 90 to 150 cm per year. Plants include shrubs and grasses. The savanna is a transitional biome between the rain forest and the desert.

Desert - temperatures range from 10-38 degrees C. Rainfall is under 25 cm per year. Plant species include xerophytes and succulents. Lizards, snakes and small mammals are common animals.

Temperate Deciduous Forest - temperature ranges from -24 to 38 degrees C. Rainfall is between 65 to 150 cm per year. Deciduous trees are common, as well as deer, bear and squirrels.

Taiga - temperatures range from -24 to 22 degrees C. Rainfall is between 35 to 40 cm per year. Taiga is located very north and very south of the equator, getting close to the poles. Plant life includes conifers and plants that can withstand harsh winters. Animals include weasels, mink, and moose.

Tundra - temperatures range from -28 to 15 degrees C. Rainfall is limited, ranging from 10 to 15 cm per year. The tundra is located even further north and south than the taiga. Common plants include lichens and mosses. Animals include polar bears and musk ox.

Polar or Permafrost - temperature ranges from -40 to 0 degrees C. It rarely gets above freezing. Rainfall is below 10 cm per year. Most water is bound up as ice. Life is limited.

Skill 9.3 Analyze factors (e.g., ecological, behavioral) that influence interrelationships among organisms.

Ecological and behavioral factors affect the interrelationships among organisms in many ways. Two important ecological factors are environmental conditions and resource availability. Important types of organismal behaviors include competitive, instinctive, territorial, and mating.

Environmental conditions, such as climate, influence organismal interrelationships by changing the dynamic of the ecosystem. Changes in climate such as moisture levels and temperature can alter the environment, changing the characteristics that are advantageous. For example, an increase in temperature will favor those organisms that can tolerate the temperature change. Thus, those organisms gain a competitive advantage. In addition, the availability of necessary resources influences interrelationships. For example, when necessary resources are scarce, interrelationships are more competitive than when resources are abundant.

Types of behavior that influence interrelationships:

Competitive – As previously mentioned, organisms compete for scarce resources. In addition, organisms compete with members of their own species for mates and territory. Many competitive behaviors involve rituals and dominance hierarchies. Rituals are symbolic activities that often settle disputes without undue harm. For example, dogs bare their teeth, erect their ears, and growl to intimidate competitors. A dominance hierarchy, or "pecking order", organizes groups of animals, simplifying interrelationships, conserving energy, and minimizing the potential for harm in a community.

Instinctive – Instinctive, or innate, behavior is common to all members of a given species and is genetically preprogrammed. Environmental differences do not affect instinctive behaviors. For example, baby birds of many types and species beg for food by raising their heads and opening their beaks.

Territorial – Many animals act aggressively to protect their territory from other animals. Animals protect territories for use in feeding, mating, and rearing of young.

Mating – Mating behaviors are very important interspecies interactions. The search for a mate with which to reproduce is an instinctive behavior. Mating interrelationships often involve ritualistic and territorial behaviors that are often competitive.

Skill 9.4   Develop a model or explanation that shows the relationships among organisms in the environment (e.g., food web, food chain, ecological pyramid).

Trophic levels are based on the feeding relationships that determine energy flow and chemical cycling.

Autotrophs are the primary producers of the ecosystem. **Producers** mainly consist of plants. **Primary consumers** are the next trophic level. The primary consumers are the herbivores that eat plants or algae. **Secondary consumers** are the carnivores that eat the primary consumers. **Tertiary consumers** eat the secondary consumer. These trophic levels may go higher depending on the ecosystem. **Decomposers** are consumers that feed off animal waste and dead organisms. This pathway of food transfer is known as the food chain.

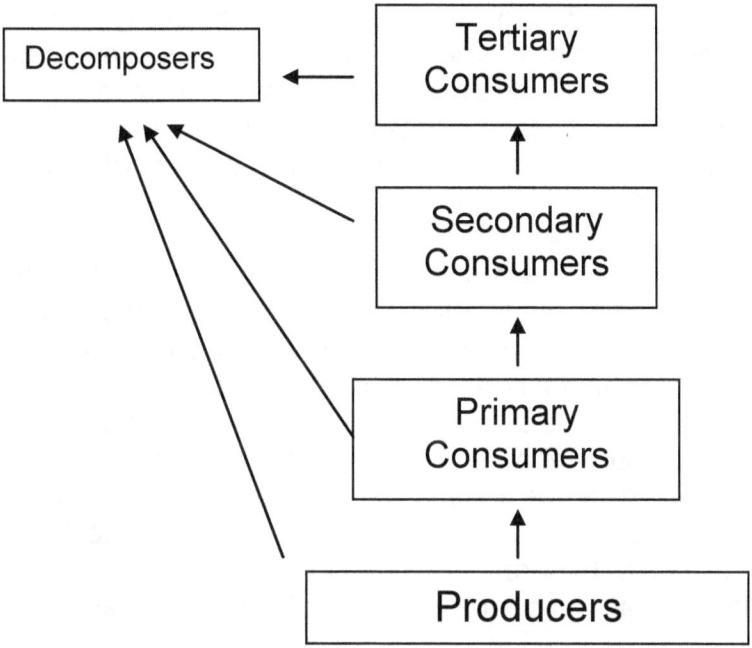

Most food chains are more elaborate, becoming food webs.

Skill 9.5 **Recognize the dynamic nature of the environment, including how communities, ecosystems, and ecoregions change over time.**

The environment is ever changing because of natural events and the actions of humans, animals, plants, and other organisms. Even the slightest changes in environmental conditions can greatly influence the function and balance of communities, ecosystems, and ecoregions. For example, subtle changes in salinity and temperature of ocean waters over time can greatly influence the range and population of certain species of fish. In addition, a slight increase in average atmospheric temperature can promote tree growth in a forest, but a corresponding increase in the viability of pathogenic bacteria can decrease the overall growth and productivity of the forest.

Another important concept in ecological change is succession. Ecological succession is the transition in the composition of species in an ecosystem, often after an ecological disturbance in the community. Primary succession begins in an environment virtually void of life, such as a volcanic island. Secondary succession occurs when a natural event disrupts an ecosystem, leaving the soil intact. An example of secondary succession is the reestablishment of a forest after destruction by a forest fire.

Factors that drive the process of succession include interspecies competition, environmental conditions, inhibition, and facilitation. In a developing ecosystem, species compete for scarce resources. The species that compete most successfully dominate. Environmental conditions, as previously discussed, influence species viability. Finally, the activities of certain species can inhibit or facilitate the growth and development of other species. Inhibition results from exploitative competition or interference competition. In facilitation, a species or group of species lays the foundation for the establishment of other, more advanced species. For example, the presence of a certain bacterial population can change the pH of the soil, allowing for the growth of different types of plants and trees.

Skill 9.6   Analyze interactions of humans with their environment.

The human population has been growing exponentially for centuries. People are living longer and healthier lives than ever before. Better health care and nutrition practices have helped in the survival of the population.

Human activity affects parts of the nutrient cycles by removing nutrients from one part of the biosphere and adding them to another. This results in nutrient depletion in one area and nutrient excess in another. This affects water systems, crops, wildlife, and humans.

Humans are responsible for the depletion of the ozone layer. This depletion is due to chemicals used for refrigeration and aerosols. The consequences of ozone depletion will be severe. Ozone protects the Earth from the majority of UV radiation. An increase of UV will promote skin cancer and unknown effects on wildlife and plants.

Humans have a tremendous impact on the world's natural resources. The world's natural water supplies are affected by human use. Waterways are major sources for recreation and freight transportation. Oil and wastes from boats and cargo ships pollute the aquatic environment. The aquatic plant and animal life is affected by this contamination.

Deforestation for urban development has resulted in the extinction or relocation of several species of plants and animals. Animals are forced to leave their forest homes or perish amongst the destruction. The number of plant and animal species that have become extinct due to deforestation is unknown. Scientists have only identified a fraction of the species on Earth. It is known that if the destruction of natural resources continues, there may be no plants or animals successfully reproducing in the wild.

Humans are continuously searching for new places to form communities. This encroachment on the environment leads to the destruction of wildlife communities. Conservationists focus on endangered species, but the primary focus should be on protecting the entire biome. If a biome becomes extinct, the wildlife dies or invades another biome.

Preservations established by the government aim at protecting small parts of biomes. While beneficial in the conservation of a few areas, the majority of the environment is still unprotected.

Skill 9.7 Explain the functions and applications of the instruments and technologies used to study the life sciences at the organism and ecosystem levels.

Scientists use a variety of tools and technologies to perform tests, collect and display data, and analyze relationships at the organismal and ecosystem level. Examples of commonly used tools include computer-linked probes, computerized tracking devices, computer models and databases, and spreadsheets.

Scientists use computer-linked probes to measure various environmental factors including temperature, dissolved oxygen, pH, ionic concentration, and pressure. The advantage of computer-linked probes, as compared to more traditional observational tools, is that the probes automatically gather data and present it in an accessible format. This property of computer-linked probes eliminates the need for constant human observation and manipulation.

Scientists use computerized tracking devices to study the behavior of animals in an ecosystem. Scientists can implant computer chips on animals to track movement and migration, population changes, and general behavioral characteristics.

Computer models allow scientists to use data and information they collect in the field to make predictions and projections about the future of ecosystems and organisms. Because ecosystems are large and change very slowly, direct observation is not a suitable strategy for ecological studies. For example, while a scientist cannot reasonably expect to observe and gather data over an entire ecosystem, she can collect samples and use computer databases and models to make projections about the ecosystem as a whole.

Finally, scientists use spreadsheets to organize, analyze, and display data. For example, conservation ecologists use spreadsheets to model population growth and development, apply sampling techniques, and create statistical distributions to analyze relationships. Spreadsheet use simplifies data collection and manipulation and allows the presentation of data in a logical and understandable format.

SUBAREA III. **PHYSICAL SCIENCE**

COMPETENCY 10.0 UNDERSTAND AND APPLY KNOWLEDGE OF THE NATURE AND PROPERTIES OF ENERGY IN ITS VARIOUS FORMS.

Skill 10.1 Describe the characteristics of and relationships among thermal, acoustical, radiant, electrical, chemical, mechanical, and nuclear energies through conceptual questions.

Thermal energy is the total internal energy of objects created by the vibration and movement of atoms and molecules. Heat is the transfer of thermal energy.

Acoustical energy, or sound energy, is the movement of energy through an object in waves. Energy that forces an object to vibrate creates sound.

Radiant energy is the energy of electromagnetic waves. Light, visible and otherwise, is an example of radiant energy.

Electrical energy is the movement of electrical charges in an electromagnetic field. Examples of electrical energy are electricity and lightning.

Chemical energy is the energy stored in the chemical bonds of molecules. For example, the energy derived from gasoline is chemical energy.

Mechanical energy is the potential and kinetic energy of a mechanical system. Rolling balls, car engines, and body parts in motion exemplify mechanical energy.

Nuclear energy is the energy present in the nucleus of atoms. Division, combination, or collision of nuclei release nuclear energy.

Skill 10.2 Analyze the processes by which energy is exchanged or transformed through conceptual questions.

Interacting objects in the universe constantly exchange and transform energy. Total energy remains the same, but the form of the energy readily changes. Energy often changes from kinetic (motion) to potential (stored) or potential to kinetic. In reality, available energy, energy that is easily utilized, is rarely conserved in energy transformations. Heat energy is an example of relatively "useless" energy often generated during energy transformations. Exothermic reactions release heat and endothermic reactions require heat energy to proceed. For example, the human body is notoriously inefficient in converting chemical energy from food into mechanical energy. The digestion of food is exothermic and produces substantial heat energy.

Skill 10.3 Apply the three laws of thermodynamics to explain energy transformations, including basic algebraic problem solving.

The three laws of thermodynamics are as follows:

1. The total amount of energy in the universe is constant, energy cannot be created or destroyed, but can merely change form.

> Equation:
> $\Delta E = Q + W$
> Change in energy = (Heat energy entering or leaving) + (work done)

2. In energy transformations, entropy (disorder) increases and useful energy is lost (as heat).

> Equation:
> $\Delta S = \Delta Q/T$
> Change in entropy = (Heat transfer) / (Temperature)

3. As the temperature of a system approaches absolute zero, entropy (disorder) approaches a constant.

Sample Problems:
1. A car engine burns gasoline to power the car. An amount of gasoline containing 2000J of stored chemical energy produced 1500J of mechanical energy to power the engine. How much heat energy did the engine release?

Solution:
$\Delta E = Q + W$	the first law of thermodynamics
2000J = Q + 1500J	apply the first law
Q (work) = 500J	

2. 18200J of heat leaks out of a hot oven. The temperature of the room is 25°C (298K). What is the increase in entropy resulting from this heat transfer?

Solution:
$\Delta S = \Delta Q/T$	the second law of thermodynamics
ΔS = 18200J / 298K	apply the second law
= 61.1 J/K	solve

Skill 10.4 Apply the principle of conservation as it applies to energy through conceptual questions and solving basic algebraic problems.

The law of conservation of energy states that energy is neither created nor destroyed. Thus, energy changes form when energy transactions occur in nature. Because the total energy in the universe is constant, energy continually transitions between forms. For example, an engine burns gasoline converting the chemical energy of the gasoline into mechanical energy, a plant converts radiant energy of the sun into chemical energy found in glucose, or a battery converts chemical energy into electrical energy.

COMPETENCY 11.0 UNDERSTAND AND APPLY KNOWLEDGE OF THE STRUCTURE AND PROPERTIES OF MATTER.

Skill 11.1 Describe the nuclear and atomic structure of matter, including the three basic parts of the atom.

An **atom** is a nucleus surrounded by a cloud with moving electrons.

The **nucleus** is the center of the atom. The positive particles inside the nucleus are called **protons.** The mass of a proton is about 2,000 times that of the mass of an electron. The number of protons in the nucleus of an atom is called the **atomic number**. All atoms of the same element have the same atomic number.

Neutrons are another type of particle in the nucleus. Neutrons and protons have about the same mass, but neutrons have no charge. Neutrons were discovered because scientists observed that not all atoms in neon gas have the same mass. They had identified isotopes. **Isotopes** of an element have the same number of protons in the nucleus, but have different masses. Neutrons explain the difference in mass. They have mass but no charge.

The mass of matter is measured against a standard mass such as the gram. Scientists measure the mass of an atom by comparing it to that of a standard atom. The result is relative mass. The **relative mass** of an atom is its mass expressed in terms of the mass of the standard atom. The isotope of the element carbon is the standard atom. It has six (6) neutrons and is called carbon-12. It is assigned a mass of 12 atomic mass units (amu). Therefore, the **atomic mass unit (amu)** is the standard unit for measuring the mass of an atom. It is equal to the mass of a carbon atom.

The **mass number** of an atom is the sum of its protons and neutrons. In any element, there is a mixture of isotopes, some having slightly more or slightly fewer protons and neutrons. The **atomic mass** of an element is an average of the mass numbers of its atoms.

The following table summarizes the terms used to describe atomic nuclei:

Term	Example	Meaning	Characteristic
Atomic Number	# protons (p)	same for all atoms of a given element	Carbon (C) atomic number = 6 (6p)
Mass number	# protons + # neutrons (p + n)	changes for different isotopes of an element	C-12 (6p + 6n) C-13 (6p + 7n)
Atomic mass	average mass of the atoms of the element	usually not a whole number	atomic mass of carbon equals 12.011

Each atom has an equal number of electrons (negative) and protons (positive). Therefore, atoms are neutral. Electrons orbiting the nucleus occupy energy levels that are arranged in order and the electrons tend to occupy the lowest energy level available. A **stable electron arrangement** is an atom that has all of its electrons in the lowest possible energy levels.

Each energy level holds a maximum number of electrons. However, an atom with more than one level does not hold more than 8 electrons in its outermost shell.

Level	Name	Max. # of Electrons
First	K shell	2
Second	L shell	8
Third	M shell	18
Fourth	N shell	32

This can help explain why chemical reactions occur. Atoms react with each other when their outer levels are unfilled. When atoms either exchange or share electrons with each other, these energy levels become filled and the atom becomes more stable.

As an electron gains energy, it moves from one energy level to a higher energy level. The electron can not leave one level until it has enough energy to reach the next level. **Excited electrons** are electrons that have absorbed energy and have moved farther from the nucleus.

Electrons can also lose energy. When they do, they fall to a lower level. However, they can only fall to the lowest level that has room for them. This explains why atoms do not collapse.

Skill 11.2   Analyze the properties of materials in relation to their chemical or physical structures (e.g., periodic table trends, relationships, and properties) and evaluate uses of the materials based on their properties.

The **periodic table of elements** is an arrangement of the elements in rows and columns so that it is easy to locate elements with similar properties. The elements of the modern periodic table are arranged in numerical order by atomic number.

The **periods** are the rows down the left side of the table. They are called first period, second period, etc. The columns of the periodic table are called **groups**, or **families.** Elements in a family have similar properties.

There are three types of elements that are grouped by color: metals, nonmetals, and metalloids.

Element Key
Atomic
Number

↓

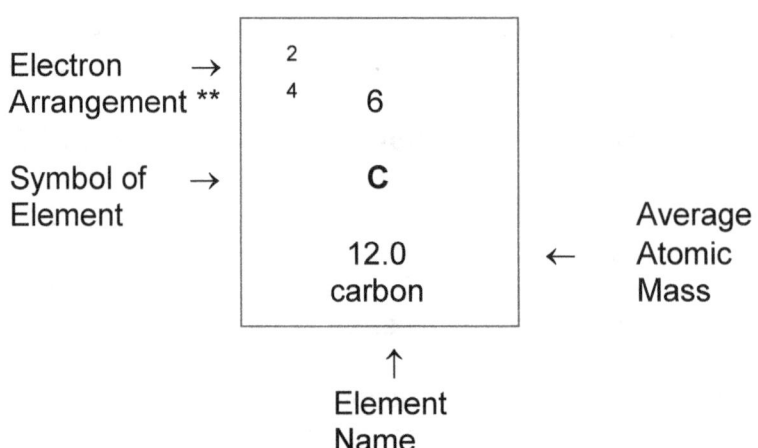

** Number of electrons on each level. Top number represents the innermost level.

The periodic table arranges metals into families with similar properties. The periodic table has its columns marked IA - VIIIA. These are the traditional group numbers. Arabic numbers 1 - 18 are also used, as suggested by the Union of Physicists and Chemists. The Arabic numerals will be used in this text.

Metals:

With the exception of hydrogen, all elements in Group 1 are **alkali metals**. These metals are shiny, softer, and less dense, and the most chemically active.

Group 2 metals are the **alkaline earth metals.** They are harder, denser, have higher melting points, and are chemically active.

The **transition elements** can be found by finding the periods (rows) from 4 to 7 under the groups (columns) 3 - 12. They are metals that do not show a range of properties as you move across the chart. They are hard and have high melting points. Compounds of these elements are colorful, such as silver, gold, and mercury.

Elements can be combined to make metallic objects. An **alloy** is a mixture of two or more elements having properties of metals. The elements do not have to be all metals. For instance, steel is made up of the metal iron and the non-metal carbon.

Nonmetals:

Nonmetals are not as easy to recognize as metals because they do not always share physical properties. However, in general the properties of nonmetals are the opposite of metals. They are not shiny, are brittle, and are not good conductors of heat and electricity.

Nonmetals are solids, gases, and one liquid (bromine).

Nonmetals have four to eight electrons in their outermost energy levels and tend to attract electrons to their outer energy levels. As a result, the outer levels usually are filled with eight electrons. This difference in the number of electrons is what caused the differences between metals and nonmetals. The outstanding chemical property of nonmetals is that react with metals.

The **halogens** can be found in Group 17. Halogens combine readily with metals to form salts. Table salt, fluoride toothpaste, and bleach all have an element from the halogen family.

The **Noble Gases** got their name from the fact that they did not react chemically with other elements, much like the nobility did not mix with the masses. These gases (found in Group 18) will only combine with other elements under very specific conditions. They are **inert** (inactive).

In recent years, scientists have found this to be only generally true, since chemists have been able to prepare compounds of krypton and xenon.

Metalloids:

Metalloids have properties in between metals and nonmetals. They can be found in Groups 13 - 16, but do not occupy the entire group. They are arranged in stair steps across the groups.

Physical Properties:
1. All are solids having the appearance of metals.
2. All are white or gray, but not shiny.
3. They will conduct electricity, but not as well as a metal.

Chemical Properties:
1. Have some characteristics of metals and nonmetals.
2. Properties do not follow patterns like metals and nonmetals. Each must be studied individually.

Boron is the first element in Group 13. It is a poor conductor of electricity at low temperatures. However, increase its temperature and it becomes a good conductor. By comparison, metals, which are good conductors, lose their ability as they are heated. It is because of this property that boron is so useful. Boron is a semiconductor. **Semiconductors** are used in electrical devices that have to function at temperatures too high for metals.

Silicon is the second element in Group 14. It is also a semiconductor and is found in great abundance in the earth's crust. Sand is made of a silicon compound, silicon dioxide. Silicon is also used in the manufacture of glass and cement.

Skill 11.3 Apply the principle of conservation as it applies to mass and charge through conceptual questions.

The principle of conservation states that certain measurable properties of an isolated system remain constant despite changes in the system. Two important principles of conservation are the conservation of mass and charge.

The principle of conservation of mass states that the total mass of a system is constant. Examples of conservation in mass in nature include the burning of wood, rusting of iron, and phase changes of matter. When wood burns, the total mass of the products, such as soot, ash, and gases, equals the mass of the wood and the oxygen that reacts with it. When iron reacts with oxygen, rust forms. The total mass of the iron-rust complex does not change. Finally, when matter changes phase, mass remains constant. Thus, when a glacier melts due to atmospheric warming, the mass of liquid water formed is equal to the mass of the glacier.

The principle of conservation of charge states that the total electrical charge of a closed system is constant. Thus, in chemical reactions and interactions of charged objects, the total charge does not change. Chemical reactions and the interaction of charged molecules are essential and common processes in living organisms and systems.

Skill 11.4 Analyze bonding and chemical, atomic, and nuclear reactions (including endothermic and exothermic reactions) in natural and man-made systems and apply basic stoichiometric principles.

Chemical reactions are the interactions of substances resulting in chemical change and change in energy. Chemical reactions involve changes in electron motion and the breaking and forming of chemical bonds. Reactants are the original substances that interact to form distinct products. Endothermic chemical reactions consume energy while exothermic chemical reactions release energy with product formation. Chemical reactions occur continually in nature and are also induced by man for many purposes.

Nuclear reactions, or **atomic reactions**, are reactions that change the composition, energy, or structure of atomic nuclei. Nuclear reactions change the number of protons and neutrons in the nucleus. The two main types of nuclear reactions are fission (splitting of nuclei) and fusion (joining of nuclei). Fusion reactions are exothermic, releasing heat energy. Fission reactions are endothermic, absorbing heat energy. Fission of large nuclei (e.g. uranium) releases energy because the products of fission undergo further fusion reactions. Fission and fusion reactions can occur naturally, but are most recognized as man-made events. Particle acceleration and bombardment with neutrons are two methods of inducing nuclear reactions.

Stoichiometry is the calculation of quantitative relationships between reactants and products in chemical reactions. Scientists use stoichiometry to balance chemical equations, make conversions between units of measurement (e.g. grams to moles), and determine the correct amount of reactants to use in chemical reactions.

Example:

The reaction of iron (Fe) and hydrochloric acid (HCl) produces H_2 and $FeCl_2$. Determine the amount of HCl required to react with 200g of Fe.

$Fe + HCl = H_2 + FeCl_2$
$Fe + 2HCl = H_2 + FeCl_2$ Balance equation (equal number of atoms on each side)

$$\frac{200 g\ Fe}{1} \cdot \frac{1\ mol\ Fe}{55.8 g\ Fe} \cdot \frac{2\ mol\ HCl}{1\ mol\ Fe} \cdot \frac{36.5 g\ HCl}{1\ mol\ HCl}$$ Perform stoichiometric calculations

= 262 g of HCl required to react completely with 200 g Fe   Solution

Skill 11.5 **Apply kinetic theory to explain interactions of energy with matter, including conceptual questions on changes in state.**

The kinetic theory states that matter consists of molecules, possessing kinetic energies, in continual random motion. The state of matter (solid, liquid, or gas) depends on the speed of the molecules and the amount of kinetic energy the molecules possess. The molecules of solid matter merely vibrate allowing strong intermolecular forces to hold the molecules in place. The molecules of liquid matter move freely and quickly throughout the body and the molecules of gaseous matter move randomly and at high speeds.

Matter changes state when energy is added or taken away. The addition of energy, usually in the form of heat, increases the speed and kinetic energy of the component molecules. Faster moving molecules more readily overcome the intermolecular attractions that maintain the form of solids and liquids. In conclusion, as the speed of molecules increases, matter changes state from solid to liquid to gas (melting and evaporation).

As matter loses heat energy to the environment, the speed of the component molecules decrease. Intermolecular forces have greater impact on slower moving molecules. Thus, as the speed of molecules decrease, matter changes from gas to liquid to solid (condensation and freezing).

Skill 11.6 **Explain the functions and applications of the instruments and technologies used to study matter and energy.**

Scientists utilize various instruments and technologies to study matter and energy. Commonly used instruments include spectrometers, basic measuring devices, thermometers, and calorimeters.

Spectroscopy is the study of absorption and emission of energy of different frequencies by molecules and atoms. The spectrometer is the instrument used in spectroscopy. Because different molecules and atoms have different spectroscopic properties, spectroscopy helps scientists determine the molecular composition of matter.

Basic devices, like scales and rulers, measure the physical properties of matter. Scientists often measure the size, volume, and mass of different forms of matter.

Thermometers and calorimeters measure the energy exchanged in chemical reactions. Temperature change during chemical reactions is indicative of the flow of energy into or out of a system of reactants and products.

COMPETENCY 12.0 UNDERSTAND AND APPLY KNOWLEDGE OF FORCES AND MOTION.

Skill 12.1 Demonstrate an understanding of the concepts and interrelationships of position, time, velocity, and acceleration through conceptual questions, algebra-based kinematics, and graphical analysis.

Position is relative. Your position in the center of the room is relative to the size of the room. A car's position on a track is relative to the determination of the track's start and end points. The velocity of an object can be defined as its speed in a particular direction. Both speed and direction are required to velocity. The symbol for velocity is v. It is the measurement of the rate of change of displacement from a fixed point. Acceleration (symbol: a) is defined as the rate of change (with respect to time) of velocity. Acceleration is measured in units of length/time², usually meters/second^2. To accelerate an object is to change its velocity, which is accomplished by altering either its speed or direction with respect to time. Acceleration has to do with changing how fast an object is moving. If an object is not changing its velocity, then the object is not accelerating. If an object is moving northward at a velocity of 10 m/s at 1 sec, and 20 m/s one second later (t= 2 sec), it is accelerating.

Skill 12.2 Demonstrate an understanding of the concepts and interrelationships of force (including gravity and friction), inertia, work, power, energy, and momentum.

Entropy is the measure of how much energy or heat is available for work. Work occurs only when heat is transferred from hot to cooler objects. Once this is done, no more work can be extracted. The energy is still being conserved, but is not available for work as long as the objects are the same temperature. Theory has it that, eventually, all things in the universe will reach the same temperature. If this happens, energy will no longer be usable.

Forces

Dynamics is the study of the relationship between motion and the forces affecting motion. **Force** causes motion.

Mass and weight are not the same quantities. An object's **mass** gives it a reluctance to change its current state of motion. It is also the measure of an object's resistance to acceleration. The force that the earth's gravity exerts on an object with a specific mass is called the object's weight on earth. Weight is a force that is measured in Newtons. Weight (W) = mass times acceleration due to gravity (**W = mg**). To illustrate the difference between mass and weight, picture two rocks of equal mass on a balance scale. If the scale is balanced in one place, it will be balanced everywhere, regardless of the gravitational field.

However, the weight of the stones would vary on a spring scale, depending upon the gravitational field. In other words, the stones would be balanced both on earth and on the moon. However, the weight of the stones would be greater on earth than on the moon.

Newton's laws of motion:

Newton's first law of motion is also called the law of inertia. It states that an object at rest will remain at rest and an object in motion will remain in motion at a constant velocity unless acted upon by an external force.

Newton's second law of motion states that if a net force acts on an object, it will cause the acceleration of the object. The relationship between force and motion is Force equals mass times acceleration. **(F = ma).**

Newton's third law states that for every action there is an equal and opposite reaction. Therefore, if an object exerts a force on another object, that second object exerts an equal and opposite force on the first.

Surfaces that touch each other have a certain resistance to motion. This resistance is **friction.**
1. The materials that make up the surfaces will determine the magnitude of the frictional force.
2. The frictional force is independent of the area of contact between the two surfaces.
3. The direction of the frictional force is opposite to the direction of motion.
4. The frictional force is proportional to the normal force between the two surfaces in contact.

Static friction describes the force of friction of two surfaces that are in contact but do not have any motion relative to each other, such as a block sitting on an inclined plane. **Kinetic friction** describes the force of friction of two surfaces in contact with each other when there is relative motion between the surfaces.

When an object moves in a circular path, a force must be directed toward the center of the circle in order to keep the motion going. This constraining force is called **centripetal force**. Gravity is the centripetal force that keeps a satellite circling the earth.

Push and pull –Pushing a volleyball or pulling a bowstring applies muscular force when the muscles expand and contract. Elastic force is when any object returns to its original shape (for example, when a bow is released).

Rubbing – Friction opposes the motion of one surface past another. Friction is common when slowing down a car or sledding down a hill.

Pull of gravity – is a force of attraction between two objects. Gravity questions can be raised not only on earth but also between planets and even black hole discussions.

Forces on objects at rest – The formula F= m/a is shorthand for force equals mass over acceleration. An object will not move unless the force is strong enough to move the mass. Also, there can be opposing forces holding the object in place. For instance, a boat may want to be forced by the currents to drift away but an equal and opposite force is a rope holding it to a dock.

Forces on a moving object - Overcoming inertia is the tendency of any object to oppose a change in motion. An object at rest tends to stay at rest. An object that is moving tends to keep moving.

Inertia and circular motion – The centripetal force is provided by the high banking of the curved road and by friction between the wheels and the road. This inward force that keeps an object moving in a circle is called centripetal force.

Simple machines include the following:

1. Inclined plane
2. Lever
3. Wheel and axle
4. Pulley

Compound machines are two or more simple machines working together. A wheelbarrow is an example of a complex machine. It uses a lever and a wheel and axle. Machines of all types ease workload by changing the size or direction of an applied force. The amount of effort saved when using simple or complex machines is called mechanical advantage or (MA).

Work is done on an object when an applied force moves through a distance.

Power is the work done divided by the amount of time that it took to do it. (Power = Work / time)

Skill 12.3 Describe and predict the motions of bodies in one and two dimensions in inertial and accelerated frames of reference in a physical system, including projectile motion but excluding circular motion.

The science of describing the motion of bodies is known as **kinematics**. The motion of bodies is described using words, diagrams, numbers, graphs, and equations.

The following words are used to describe motion: vectors, scalars, distance, displacement, speed, velocity, and acceleration.

The two categories of mathematical quantities that are used to describe the motion of objects are scalars and vectors. **Scalars** are quantities that are fully described by magnitude alone. Examples of scalars are 5m and 20 degrees Celsius. **Vectors** are quantities that are fully described by magnitude and direction. Examples of vectors are 30m/sec, and 5 miles north.

Distance is a scalar quantity that refers to how much ground an object has covered while moving. **Displacement** is a vector quantity that refers to the object's change in position.

Example:

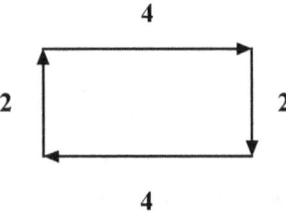

Jamie walked 2 miles north, 4 miles east, 2 miles south, and then 4 miles west. In terms of distance, she walked 12 miles. However, there is no displacement because the directions cancelled each other out, and she returned to her starting position.

Speed is a scalar quantity that refers to how fast an object is moving (ex. the car was traveling 60 mi./hr). **Velocity** is a vector quantity that refers to the rate at which an object changes its position. In other words, velocity is speed with direction (ex. the car was traveling 60 mi./hr east).

$$\text{Average speed} = \frac{\text{Distance traveled}}{\text{Time of travel}}$$

$$v = \frac{d}{t}$$

$$\text{Average velocity} = \frac{\Delta \text{position}}{\text{time}} = \frac{\text{displacement}}{\text{time}}$$

Instantaneous Speed - speed at any given instant in time.

Average Speed - average of all instantaneous speeds, found simply by a distance/time ratio.

Acceleration is a vector quantity defined as the rate at which an object changes its velocity.

$$a = \frac{\Delta velocity}{time} = \frac{v_f - v_i}{t}$$ where *f* represents the final velocity and *i*

represents the initial velocity

Since acceleration is a vector quantity, it always has a direction associated with it. The direction of the acceleration vector depends on

- whether the object is speeding up or slowing down
- whether the object is moving in the positive or negative direction.

Newton's Three Laws of Motion:

First Law: An object at rest tends to stay at rest and an object in motion tends to stay in motion with the same speed and in the same direction unless acted upon by an unbalanced force, for example, when riding on a descending elevator that suddenly stops, blood rushes from your head to your feet. **Inertia** is the resistance an object has to a change in its state of motion.

Second Law: The acceleration of an object depends directly upon the net force acting upon the object, and inversely upon the mass of the object. As the net force increases, so will the object's acceleration. However, as the mass of the object increases, its acceleration will decrease.

$$F_{net} = m * a$$

Third Law: For every action, there is an equal and opposite reaction, for example, when a bird is flying, the motion of its wings pushes air downward; the air reacts by pushing the bird upward.

Projectile Motion

By definition, a **projectile** has only one force acting upon it – the force of gravity. Gravity influences the vertical motion of the projectile, causing vertical acceleration. The horizontal motion of the projectile is the result of the tendency of any object in motion to remain in motion at constant velocity. (Remember, there are no horizontal forces acting upon the projectile. By definition, gravity is the only force acting upon the projectile.)

Projectiles travel with a parabolic trajectory due to the fact that the downward force of gravity accelerates them downward from their otherwise straight-line trajectory. Gravity affects the vertical motion, not the horizontal motion, of the projectile. Gravity causes a downward displacement from the position that the object would be in if there were no gravity.

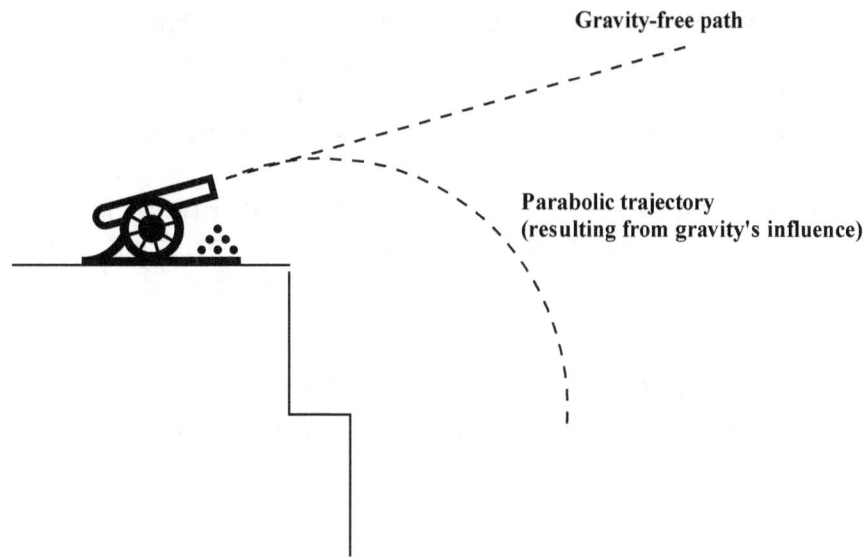

Skill 12.4 Analyze and predict motions and interactions of bodies involving forces within the context of conservation of energy and/or momentum through conceptual questions and algebra-based problem solving.

The Law of **Conservation of Energy** states that energy may neither be created nor destroyed. Therefore, the sum of all energies in the system is a constant.

Example:

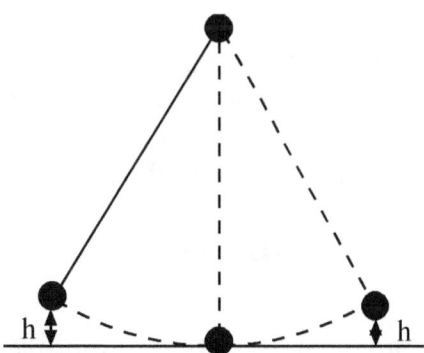

The formula to calculate the potential energy is PE = mgh.

The mass of the ball = 20kg
The height, h = 0.4m
The acceleration due to gravity, g = 9.8 m/s^2

PE = mgh
PE = 20(.4)(9.8)
PE = 78.4J (Joules, units of energy)

The position of the ball on the left is where the Potential Energy (PE) = 78.4J resides while the Kinetic Energy (KE) = 0. As the ball is approaching the center position, the PE is decreasing while the KE is increasing. At exactly halfway between the left and center positions, the PE = KE.

The center position of the ball is where the Kinetic Energy is at its maximum while the Potential Energy (PE) = 0. At this point, theoretically, the entire PE has transformed into KE. Now the KE = 78.4J while the PE = 0.

The right position of the ball is where the Potential Energy (PE) is once again at its maximum and the Kinetic Energy (KE) = 0.

We can now say that:

$$PE + KE = 0$$
$$PE = -KE$$

The sum of PE and KE is the **total mechanical energy:**

$$\text{Total Mechanical Energy} = PE + KE$$

The law of **momentum conservation** can be stated as follows: For a collision occurring between object 1 and object 2 in an isolated system, the total momentum of the two objects before the collision is equal to the total momentum of the two objects after the collision. That is, the momentum lost by object 1 is equal to the momentum gained by object 2.

Example:

A 90 kg soccer player moving west at 4 m/s collides with an 80 kg soccer player moving east at 5 m/s. After the collision, both players move east at 3 m/s. Draw a vector diagram in which the before and after collision momenta of each player is represented by a momentum vector.

BEFORE

400 kg m/s →   ← 360 kg m/s

AFTER
combined unit

40 kg m/s →

The combined momentum before the collision is +400 kg m/s + (- 360 kg m/s) or 40 kg m/s. According to the law of conservation of momentum, the combined momentum after a collision must be equal to the combined momentum before the collision.

Skill 12.5   Describe the effects of gravitational and nuclear forces in real-life situations through conceptual questions.

Gravitational and the two nuclear forces make up three of the four fundamental forces of nature, with electromagnetic force being the fourth.

Gravitational force is defined as the force of attraction between all masses in the universe. Every object exerts gravitational force on every other object. This force depends on the masses of the objects and the distance between them. The gravitational force between any two masses is given by Newton's law of universal gravitation, which states that the force is inversely proportional to the square of the distance between the masses. Near the surface of the Earth, the acceleration of an object due to gravity is independent of the mass of the object and therefore constant.

It is the gravitational attraction of the Earth that gives weight to objects with mass and causes them to fall to the ground when dropped. Gravitation is responsible for the existence of the objects in our solar system. Without it, the celestial bodies would not be held together. It keeps the Earth and all the other planets in orbit around the sun, keeps the moon in orbit around the Earth, and causes the formation of the tides.

Examples of mechanisms that utilize gravitation to some degree are intravenous drips and water towers where the height difference provides a pressure differential in the liquid. The gravitational potential energy of water is also used to generate hydroelectricity. Pendulum clocks depend upon gravity to regulate time.

There are two types of nuclear forces: strong and weak. The strong force is an interaction that binds protons and neutrons together in atomic nuclei. The strong force only acts on elementary particles directly, but is observed between hadrons (subatomic particles) as the nuclear force.

The weak force is an interaction between elementary particles involving neutrinos or antineutrinos. Its most familiar effects are beta decay and the associated radioactivity.

Skill 12.6 Explain the functions and applications of the instruments and technologies used to study force and motion in everyday life.

The **speedometer** in a car indicates how fast the car is going at any moment (instantaneous speed).

Tachometers measure the speed of rotation of a shaft or disk. In an automobile, this assists the driver in choosing gear settings with a manual transmission. Tachometers are also used in medicine to measure blood flow.

Gravity gradiometers are used in petroleum exploration to determine areas of higher or lower density in the earth's crust.

An **accelerometer** is a device for measuring acceleration. Accelerometers are used along with gyroscopes in inertial guidance systems for rocket programs. One of the most common uses for accelerometers is in airbag deployment systems in automobiles. The accelerometers are used to detect rapid deceleration of the vehicle to determine when a collision has occurred and the severity of the collision. Research is currently being done on the use of accelerometers to improve Global Positioning Systems (GPS). An accelerometer can infer position in places such as tunnels where the GPS cannot detect it. They are incorporated into Tablet PCs to align the screen based upon the direction in which the PC is being held and in many laptop computers to detect falling and protect the data on the hard drive. Accelerometers are incorporated into sports watches to indicate speed and distance (useful to runners).

A **gravimeter** is a device used to measure the local gravitational field. They are much more sensitive than accelerometers. Measurements of the surface gravity of the earth are part of geophysical analysis, which includes the study of earthquakes.

Weighing scales, such as spring scales, are sometimes used to measure force rather than mass or weight.

COMPETENCY 13.0 UNDERSTAND AND APPLY KNOWLEDGE OF ELECTRICITY, MAGNETISM, AND WAVES.

Skill 13.1 Recognize the nature and properties of electricity and magnetism, including static charge, moving charge, basic RC circuits, fields, conductors, and insulators.

An **electric circuit** is a path along which electrons flow. A simple circuit can be created with a dry cell, wire, a bell, or a light bulb. When all are connected, the electrons flow from the negative terminal, through the wire to the device and back to the positive terminal of the dry cell. If there are no breaks in the circuit, the device will work. The circuit is closed. Any break in the flow will create an open circuit and cause the device to shut off.

The device (bell, bulb) is an example of a **load**. A load is a device that uses energy. Suppose that you also add a buzzer so that the bell rings when you press the buzzer button. The buzzer is acting as a **switch**. A switch is a device that opens or closes a circuit. Pressing the buzzer makes the connection complete and the bell rings. When the buzzer is not engaged, the circuit is open and the bell is silent.

A **series circuit** is one where the electrons have only one path along which they can move. When one load in a series circuit goes out, the circuit is open. An example of this is a set of Christmas tree lights that is missing a bulb. None of the bulbs will work.

A **parallel circuit** is one where the electrons have more than one path to move along. If a load goes out in a parallel circuit, the other load will still work because the electrons can still find a way to continue moving along the path.

When an electron goes through a load, it does work and therefore loses some of its energy. The measure of how much energy is lost is called the **potential difference**. The potential difference between two points is the work needed to move a charge from one point to another.

Potential difference is measured in a unit called the volt. **Voltage** is potential difference. The higher the voltage, the more energy the electrons have. This energy is measured by a device called a voltmeter. To use a voltmeter, place it in a circuit parallel with the load you are measuring.

Current is the number of electrons per second that flow past a point in a circuit. Current is measured with a device called an ammeter. To use an ammeter, put it in series with the load you are measuring.

As electrons flow through a wire, they lose potential energy. Some is changed into heat energy because of resistance. **Resistance** is the ability of the material to oppose the flow of electrons through it. All substances have some resistance, even if they are a good conductor such as copper. This resistance is measured in units called **ohms**. A thin wire will have more resistance than a thick one because it will have less room for electrons to travel. In a thicker wire, there will be more possible paths for the electrons to flow. Resistance also depends upon the length of the wire. The longer the wire, the more resistance it will have. Potential difference, resistance, and current form a relationship know as **Ohm's Law**. Current **(I)** is measured in amperes and is equal to potential difference **(V)** divided by resistance **(R)**.

$$I = V / R$$

If you have a wire with resistance of 5 ohms and a potential difference of 75 volts, you can calculate the current by

$$I = 75 \text{ volts} / 5 \text{ ohms}$$
$$I = 15 \text{ amperes}$$

A current of 10 or more amperes will cause a wire to get hot. 22 amperes is about the maximum for a house circuit. Anything above 25 amperes can start a fire.

Electrostatics is the study of stationary electric charges. A plastic rod that is rubbed with fur or a glass rod that is rubbed with silk will become electrically charged and will attract small pieces of paper. The charge on the plastic rod rubbed with fur is negative and the charge on glass rod rubbed with silk is positive.

Electrically charged objects share these characteristics:

1. Like charges repel one another.
2. Opposite charges attract each other.
3. Charge is conserved. A neutral object has no net change. If the plastic rod and fur are initially neutral, when the rod becomes charged by the fur a negative charge is transferred from the fur to the rod. The net negative charge on the rod is equal to the net positive charge on the fur.

Materials through which electric charges can easily flow are called **conductors**. Metals which are good conductors include silicon and boron. On the other hand, an **insulator** is a material through which electric charges do not move easily, if at all. Examples of insulators would be the nonmetal elements of the periodic table. A simple device used to indicate the existence of a positive or negative charge is called an **electroscope**. An electroscope is made up of a conducting knob and attached to it are very lightweight conducting leaves usually made of foil (gold or aluminum). When a charged object touches the knob, the leaves push away from each other because like charges repel. It is not possible to tell whether if the charge is positive or negative.

Charging by induction:

Touch the knob with a finger while a charged rod is nearby. The electrons will be repulsed and flow out of the electroscope through the hand. If the hand is removed while the charged rod remains close, the electroscope will retain the charge.

When an object is rubbed with a charged rod, the object will take on the same charge as the rod. However, charging by induction gives the object the opposite charge as that of the charged rod.

Grounding charge:

Charge can be removed from an object by connecting it to the earth through a conductor. The removal of static electricity by conduction is called **grounding**.

Skill 13.2   Recognize the nature and properties of mechanical and electromagnetic waves (e.g., frequency, source, medium, spectrum, wave-particle duality).

A mechanical wave can be defined as a disturbance that travels through a medium, moving energy from one place to another. This disturbance is also called an electrical force field. The wave is not capable of transporting energy without a medium, as in vacuum conditions. The medium is the material through which the disturbance is moving and can be thought of as a series of interacting particles. The example of a slinky wave is often used to illustrate the nature of a wave, where pressure exerted on the first coil moves through the remaining coils. A sound wave is also a mechanical wave. The frequency of a wave refers to how often the particles of the medium vibrate when a wave passes through the medium. Frequency is measured in units of cycles/second, waves/second, vibrations/second, or something/second. Another unit for frequency is the Hertz (abbreviated Hz) where 1 Hz is equivalent to 1 cycle/second. The period of a wave is the time required for a particle on a medium to make one complete vibrational cycle. Wave period is measured in units of time such as seconds, hours, days or years, and is NOT synonymous with frequency. Electromagnetic waves are both electric and magnetic in nature and are capable of traveling through a vacuum. They do not require a medium in order to transport their energy. Light, microwaves, x-rays, and TV and radio transmissions are all kinds of electromagnetic waves. They are all a wavy disturbance that repeats itself over a distance called the wavelength. Electromagnetic waves come in varying sizes and properties, by which they are organized in the electromagnetic spectrum. The electromagnetic spectrum is measured in frequency (f) in hertz and wavelength (λ) in meters. The frequency times the wavelength of every electromagnetic wave equals the speed of light (3.0×10^9 meters/second).

Roughly, the range of wavelengths of the electromagnetic spectrum is:

	f	λ
Radio waves	$10^5 - 10^{-1}$ hertz	$10^3 - 10^9$ meters
Microwaves	$10^{-1} - 10^{-3}$ hertz	$10^9 - 10^{11}$ meters
Infrared radiation	$10^{-3} - 10^{-6}$ hertz	$10^{11.2} - 10^{14.3}$ meters
Visible light	$10^{-6.2} - 10^{-6.9}$ hertz	$10^{14.3} - 10^{15}$ meters
Ultraviolet radiation	$10^{-7} - 10^{-9}$ hertz	$10^{15} - 10^{17.2}$ meters
X-Rays	$10^{-9} - 10^{-11}$ hertz	$10^{17.2} - 10^{19}$ meters
Gamma Rays	$10^{-11} - 10^{-15}$ hertz	$10^{19} - 10^{23.25}$ meters

Radio waves are used for transmitting data. Common examples are television, cell phones, and wireless computer networks. Microwaves are used to heat food and deliver Wi-Fi service. Infrared waves are utilized in night vision goggles. Visible light we are all familiar with as the human eye is most sensitive to this wavelength range. UV light causes sunburns and would be even more harmful if most of it were not captured in the Earth's ozone layer. X-rays aid us in the medical field and gamma rays are most useful in the field of astronomy.

Skill 13.3   Describe the effects and applications of electromagnetic forces in real-life situations, including electric power generation, circuit breakers, and brownouts.

Electricity can be used to change the chemical composition of a material. For instance, when electricity is passed through water, it breaks the water down into hydrogen gas and oxygen gas.

Circuit breakers in a home monitor the electric current. If there is an overload, the circuit breaker will create an open circuit, stopping the flow of electricity.

Computers can be made small enough to fit inside a plastic credit card by creating what is known as a solid state device. In this device, electrons flow through solid material such as silicon.

Resistors are used to regulate volume on a television or radio or through a dimmer switch for lights.

A bird can sit on an electrical wire without being electrocuted because the bird and the wire have about the same potential. However, if that same bird would touch two wires at the same time he would not have to worry about flying south next year.

When caught in an electrical storm, a car is a relatively safe place from lightening because of the resistance of the rubber tires. A metal building would not be safe unless there was a lightening rod that would attract the lightening and conduct it into the ground.

A brown-out occurs when there exists a condition of lower than normal power line voltage. This may be short term (minutes to hours) or long term (1/2 day or more). A power line voltage reduction of 8 - 12% is usually considered a Brown-out. Electric utilities may reduce line voltage to Brown-out levels in an effort to manage power generation and distribution. This is most likely to occur on very hot days, when most air conditioning and refrigeration equipment would be operating almost continuously. Even without purposeful intervention from the local utility company, extreme overloads (spikes) could tax the electrical system to the point where a permanent brown-out state could exist over much of the company's distribution network.

Skill 13.4 Analyze and predict the behavior of mechanical and electromagnetic waves under varying physical conditions, including basic optics, color, ray diagrams, and shadows.

The place where one medium ends and another begins is called a **boundary**, and the manner in which a wave behaves when it reaches that boundary is called **boundary behavior**. The following principles apply to boundary behavior in waves:

1) wave speed is always greater in the less dense medium
2) wavelength is always greater in the less dense medium
3) wave frequency is not changed by crossing a boundary
4) the reflected pulse becomes inverted when a wave in a less dense medium is heading towards a boundary with a more dense medium
5) the amplitude of the incident pulse is always greater than the amplitude of the reflected pulse.

For an example, we will use a rope whose left side is less dense, or thinner, than the right side of the rope.

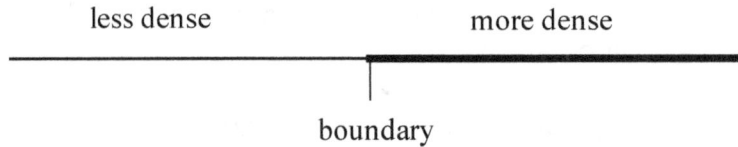

A pulse is introduced on the left end of the rope. This **incident pulse** travels right along the rope towards the boundary between the two thicknesses of rope. When the incident pulse reaches the boundary, two behaviors will occur:

1) Some of the energy will be reflected back to the left side of the boundary. This energy is known as the **reflected pulse**.
2) The rest of the energy will travel into the thicker end of the rope. This energy is referred to as the **transmitted pulse**.

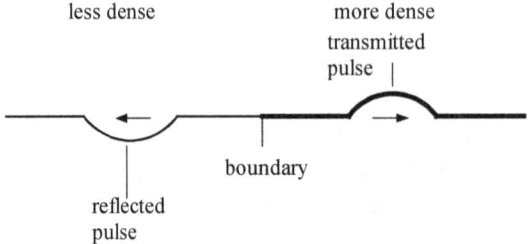

When the incident pulse travels from a denser medium to a less dense medium, the reflected pulse is not inverted.

Reflection occurs when waves bounce off a barrier. The **law of reflection** states that when a ray of light reflects off a surface, the angle of incidence is equal to the angle of reflection.

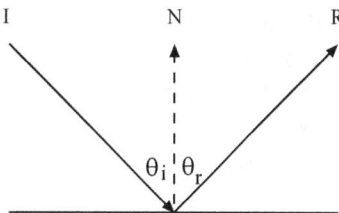

Line I represents the **incident ray**, the ray of light striking the surface. Line R is the **reflected ray**, the ray of light reflected off the surface. Line N is known as the **normal line**. It is a perpendicular line at the point of incidence that divides the angle between the incident ray and the reflected ray into two equal rays. The angle between the incident ray and the normal line is called the **angle of incidence**; the angle between the reflected ray and the normal line is called the **angle of reflection**.

Waves passing from one medium into another will undergo **refraction**, or bending. Accompanying this bending are a change in both speed and the wavelength of the waves.

In this example, light waves traveling through the air will pass through glass.

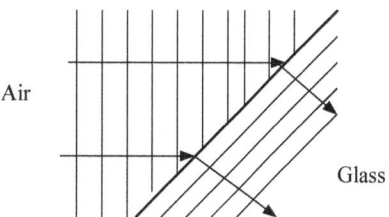

Refraction occurs only at the boundary. Once the wavefront passes across the boundary, it travels in a straight line.

Diffraction involves a change in direction of waves as they pass through an opening or around an obstacle in their path.

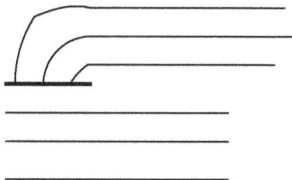

The amount of diffraction depends upon the wavelength. The amount of diffraction increases with increasing wavelength and decreases with decreasing wavelength. Sound and water waves exhibit this ability.

When we refer to light, we are usually talking about a type of electromagnetic wave that stimulates the retina of the eye, or visible light. Each individual wavelength within the spectrum of visible light represents a particular **color**. When a particular wavelength strikes the retina, we perceive that color. Visible light is sometimes referred to as ROYGBIV (red, orange, yellow, green, blue, indigo, violet). The visible light spectrum ranges from red (the longest wavelength) to violet (the shortest wavelength) with a range of wavelengths in between. If all the wavelengths strike your eye at the same time, you will see white. Conversely, when no wavelengths strike your eye, you perceive black.

A **shadow** results from the inability of light waves to diffract as sound and water waves can. An obstacle in the way of the light waves blocks the light waves, thereby creating a shadow.

TEACHER CERTIFICATION STUDY GUIDE

SUBAREA IV. EARTH SYSTEMS AND THE UNIVERSE

COMPETENCY 14.0 UNDERSTAND AND APPLY KNOWLEDGE OF EARTH'S LAND, WATER, AND ATMOSPHERIC SYSTEMS AND THE HISTORY OF EARTH.

Skill 14.1 Identify the structure and composition of Earth's land, water, and atmospheric systems and how they affect weather, erosion, fresh water, and soil.

Water is recycled throughout ecosystems. Just two percent of all the available water is fixed and held in ice or the bodies of organisms. Available water includes surface water (lakes, ocean, and rivers) and ground water (aquifers, wells). 96% of all available water is from ground water. Water is recycled through the processes of evaporation and precipitation. The water present now is the water that has been here since our atmosphere formed.

When water falls from the atmosphere it can have an erosive property. This can happen from impact alone but also from acid rain. Erosion is known for its destructive properties, but in an indirect way, it also builds by bringing materials to new locations. **Erosion** is the inclusion and transportation of surface materials by another moveable material, usually water, wind, or ice. The most important cause of erosion is running water. Streams, rivers, and tides are constantly at work removing weathered fragments of bedrock and carrying them away from their original location. A stream erodes bedrock by the grinding action of the sand, pebbles and other rock fragments. This grinding against each other is called abrasion. Streams also erode rocks by dissolving or absorbing their minerals. Limestone and marble are readily dissolved by streams.

The breaking down of rocks at or near to the earth's surface is known as **weathering**. Weathering breaks down these rocks into smaller and smaller pieces. There are two types of weathering: physical weathering and chemical weathering.

Physical weathering is the process by which rocks are broken down into smaller fragments without undergoing any change in chemical composition. Physical weathering is mainly caused by the freezing of water, the expansion of rock, and the activities of plants and animals.

Frost wedging is the cycle of daytime thawing and refreezing at night. This cycle causes large rock masses, especially the rocks exposed on mountain tops, to be broken into smaller pieces.

The peeling away of the outer layers from a rock is called exfoliation. Rounded mountain tops are called exfoliation domes and have been formed in this way.

EARTH & SPACE SCIENCE

Chemical weathering is the breaking down of rocks through changes in their chemical composition. An example would be the change of feldspar in granite to clay. Water, oxygen, and carbon dioxide are the main agents of chemical weathering. When water and carbon dioxide combine chemically, they produce a weak acid that breaks down rocks. In addition, acidic substances from factories and car exhausts dissolve in rain water forming **acid rain.** Acid rain forms predominantly from pollutant oxides in the air (usually nitrogen-based NO_x or sulfur-based SO_x), which become hydrated into their acids (nitric or sulfuric acid). When the rain falls into stone, the acids can react with metallic compounds and gradually wear the stone away.

Skill 14.2   Recognize the scope of geologic time and the continuing physical changes of Earth through time.

Geological time is divided into periods depending on the kind of life that existed at that time. These periods are grouped together into eras. The history of the Earth is calculated by studying the ages of the various layers of sedimentary rock.

Era	Period	Time	Characteristics
Cenozoic	Quaternary	1.6 million years ago to the present.	The Ice Age occurred, and human beings evolved.
	Tertiary	65-1.64 million years ago.	Mammals and birds evolved to replace the great reptiles and dinosaurs that had just become extinct. Forests gave way to grasslands, and the climate become cooler.
Mesozoic	Cretaceous	135-65 million years ago.	Reptiles and dinosaurs roamed the Earth. Most of the modern continents had split away from the large landmass, Pangaea, and many were flooded by shallow chalk seas.
	Jurassic Triassic	350-135 million years ago.	Reptiles were beginning to evolve. Pangaea started to break up. Deserts gave way to forests and swamps.
Paleozoic	**Permian** **Carboniferous**	355-250 million years ago.	Continents came together to form one big landmass, Pangaea. Forests (that formed today's coal) grew on deltas around the new mountains, and deserts formed.

	Devonian	410-355 million years ago.	Continents started moving toward each other. The first land animals, such as insects and amphibians, existed. Many fish swam in the seas.
	Silurian	510-410 million years ago.	Sea life flourished, and the first fish evolved. The earliest land plants began to grow around shorelines and estuaries.
	Ordovician		
	Cambrian	570-510 million years ago.	No life on land, but all kinds of sea animals existed.
Precambrian	**Proterozoic**	Beginning of the Earth to 570 million years ago (seven-eighths of the Earth's history).	Some sort of life existed.
	Archaean		No life.

Skill 14.3 Evaluate scientific theories about Earth's origin and history and how these theories explain contemporary living systems

The dominant scientific theory about the origin of the Universe, and consequently the Earth, is the **Big Bang Theory**. According to this theory, an atom exploded about 10 to 20 billion years ago throwing matter in all directions. Although this theory has never been proven, and probably never will be, it is supported by the fact that distant galaxies in every direction are moving away from us at great speeds.

Earth, itself, is believed to have been created 4.5 billion years ago as a solidified cloud of gases and dust left over from the creation of the sun. As millions of years passed, radioactive decay released energy that melted some of Earth's components. Over time, the heavier components sank to the center of the Earth and accumulated into the core. As the Earth cooled, a crust formed with natural depressions. Water rising from the interior of the Earth filled these depressions and formed the oceans. Slowly, the Earth acquired the appearance it has today.

The **Heterotroph Hypothesis** supposes that life on Earth evolved from **heterotrophs**, the first cells. According to this hypothesis, life began on Earth about 3.5 billion years ago. Scientists have shown that the basic molecules of life formed from lightning, ultraviolet light, and radioactivity. Over time, these molecules became more complex and developed metabolic processes, thereby becoming heterotrophs. Heterotrophs could not produce their own food and fed off organic materials. However, they released carbon dioxide that allowed for the evolution of **autotrophs**, which could produce their own food through photosynthesis. The autotrophs and heterotrophs became the dominant life forms and evolved into the diverse forms of life we see today.

Proponents of **creationism** believe that the species we currently have were created as recounted in the book of Genesis in the Bible. This retelling asserts that God created all life about 6,000 years ago in one mass creation event. However, scientific evidence casts doubt on creationism.

Evolution

The most significant evidence to support the history of evolution is fossils, which have been used to construct a fossil record. Fossils give clues as to the structure of organisms and the times at which they existed. However, there are limitations to the study of fossils, which leave huge gaps in the fossil record.

Scientists also try to relate two organisms by comparing their internal and external structures. This is called **comparative anatomy**. Comparative anatomy categorizes anatomical structures as **homologous** (features in different species that point to a common ancestor), **analogous** (structures that have superficial similarities because of similar functions, but do not point to a common ancestor), and **vestigial** (structures that have no modern function, indicating that different species diverged and evolved). Through the study of **comparative embryology**, homologous structures that do not appear in mature organisms may be found between different species in their embryological development.

There have been two basic **theories of evolution: Lamarck's and Darwin's**. Lamarck's theory that proposed that an organism can change its structure through use or disuse and that acquired traits can be inherited has been disproved.

Darwin's theory of **natural selection** is the basis of all evolutionary theory. His theory has four basic points:

1. Each species produces more offspring than can survive.
2. The individual organisms that make up a larger population are born with certain variations.
3. The overabundance of offspring creates competition for survival among individual organisms (**survival of the fittest**).
4. Variations are passed down from parent to offspring.

Points 2 and 4 form the genetic basis for evolution.

New species develop from two types of evolution: divergent and convergent. **Divergent evolution**, also known as **speciation**, is the divergence of a new species from a previous form of that species. There are two main ways in which speciation may occur: **allopatric speciation** (resulting from geographical isolation so that species cannot interbreed) and **adaptive radiation** (creation of several new species from a single parent species). **Convergent evolution** is a process whereby different species develop similar traits from inhabiting similar environments, facing similar selection pressures, and/or use parts of their bodies for similar functions. This type of evolution is only superficial. It can never result in two species being able to interbreed.

Skill 14.4 Recognize the interrelationships between living organisms and Earth's resources and evaluate the uses of Earth's resources

The region of the Earth and its atmosphere in which living things are found is known as the **biosphere**. The biosphere is made up of distinct areas called **ecosystems**, each of which has its own characteristic climate, soils, and communities of plants and animals.

The most important nonliving factors affecting an ecosystem are the chemical cycles, the water cycle, oxygen, sunlight, and the soil. The two basic chemical cycles are the carbon cycle and the nitrogen cycle. They involve the passage of these elements between the organisms and the environment.

In the **carbon cycle**, animals and plants use carbon dioxide from the air to produce glucose, which they use in respiration and other life processes. Animals consume plants, use what they can of the carbon matter and excrete the rest as waste. This waste decays into carbon dioxide. During respiration, plants and animals release carbon dioxide back into the air. The carbon used by plants and animals stays in their bodies until death, after which decay sends the organic compounds back into the Earth and carbon dioxide back into the air.

Nitrogen found in the atmosphere is generally unusable by living organisms. In the **nitrogen cycle,** nitrogen-fixing bacteria in the soil and/or the roots of legumes transform the inert nitrogen into compounds. Plants take these compounds, synthesize the twenty amino acids found in nature, and turn them into plant proteins. Animals can only synthesize eight of the amino acids. They eat the plants to produce protein from the plant's materials. Animals and plants give off nitrogen waste and death products in the form of ammonia. The ammonia will either be transformed into nitrites and nitrates by bacteria and reenter the cycle when they are taken up by plants, or be broken down by bacteria to produce inert nitrogen to be released back into the air.

Most of the Earth's water is found in the oceans and lakes. Through the **water cycle**, water evaporates into the atmosphere and condenses into clouds. Water then falls to the Earth in the form of precipitation, returning to the oceans and lakes on falling on land. Water on the land may return to the oceans and lakes as runoff or seep from the soil as groundwater.

The amount of **oxygen** available in a particular location may create competition. Oxygen is readily available to animals on land; but in order for it to be available to aquatic organisms, it must be dissolved in water.

Sunlight is also important to most organisms. Organisms on land compete for sunlight, but sunlight does not reach into the lowest depths of the ocean. Organisms in these regions must find another means of producing food.

The type of **soil** found in a particular ecosystem determines what species can live in that ecosystem. The ecosystems of the Earth consist of: temperate forests, deserts, wetlands, tropical rain forests, oceans, grasslands, rivers and lakes, mountains, towns and cities, seashores, and polar and tundra lands.

COMPETENCY 15.0 UNDERSTAND AND APPLY KNOWLEDGE OF THE DYNAMIC NATURE OF EARTH.

Skill 15.1 Analyze and explain large-scale dynamic forces, events, and processes that affect Earth's land, water, and atmospheric systems, including conceptual questions about plate tectonics, El Niño, drought, and climatic shifts.

Air masses moving toward or away from the Earth's surface are called air currents. Air moving parallel to Earth's surface is called **wind**. Weather conditions are generated by winds and air currents carrying large amounts of heat and moisture from one part of the atmosphere to another. Wind speeds are measured by instruments called anemometers.

The wind belts in each hemisphere consist of convection cells that encircle Earth like belts. There are three major wind belts on Earth (1) trade winds (2) prevailing westerlies, and (3) polar easterlies. Wind belt formation depends on the differences in air pressures that develop in the doldrums, the horse latitudes, and the polar regions. The Doldrums surround the equator. Within this belt heated air usually rises straight up into Earth's atmosphere. The Horse latitudes are regions of high barometric pressure with calm and light winds and the Polar regions contain cold dense air that sinks to the earth's surface

Winds caused by local temperature changes include sea breezes, and land breezes.

Sea breezes are caused by the unequal heating of the land and an adjacent, large body of water. Land heats up faster than water. The movement of cool ocean air toward the land is called a sea breeze. Sea breezes usually begin blowing about mid-morning; ending about sunset.

A breeze that blows from the land to the ocean or a large lake is called a **land breeze.**

Monsoons are huge wind systems that cover large geographic areas and that reverse direction seasonally. The monsoons of India and Asia are examples of these seasonal winds. They alternate wet and dry seasons. As denser cooler air over the ocean moves inland, a steady seasonal wind called a summer or wet monsoon is produced.

El Niño refers to a sequence of changes in the ocean and atmospheric circulation across the Pacific Ocean. The water around the equator is unusually hot every two to seven years. Trade winds normally blowing east to west across the equatorial latitudes, piling warm water into the western Pacific. A huge mass of heavy thunderstorms usually forms in the area and produce vast currents of rising air that displace heat poleward. This helps create the strong mid-latitude jet streams. The world's climate patterns are disrupted by this change in location of the massive cluster of thunderstorms. The West coast of America experienced a wet winter. Sacramento, California recorded 103 days of rain.

Drought
A drought is a period of time of abnormal dryness. Drought is caused by atmospheric flow. In times of normal precipitation, water vapor rises with a low pressure system, condenses, forms clouds, and precipitation occurs. In times of drought, however, areas of high pressure stall allowing the water vapor to stay close to the surface. If the water vapor does not rise, it cannot condense to form clouds and thus precipitation is low. High pressure systems stall because of upper atmospheric conditions and the jet stream. La Niña, a cooling of the water in the Pacific, can also cause drought. If the water in the Pacific is cooler, less air rises over the Pacific. The water vapor does not rise, therefore clouds and precipitation do not form. Periods of drought do not have to be periods of time with complete lack of precipitation. Times of minimal precipitation cause drought as well. If the input of precipitation in the system is less than the amount of water going out through plant and human use, drought occurs.

Climatic shift
Our planet goes through normal periods of climatic shift. One reason for climate shift is change in the sun's output of energy. The sun goes through a pattern of sunspot activity over a period of 11 years. It has been noted that the amount of sunspots may have an effect on the solar output, thereby having an effect on the earth's climate. In the mid 1600's to early 1700's, there was a period of time referred to as the Maunder Minimum. During this time, astronomers only observed about 50 sunspots as opposed to the normal 40,000 to 50,000 sunspots. The Maunder Minimum corresponded to a period of exceedingly cold winters over Europe and North America. This period of time has been dubbed the "Little Ice Age".

Climate change has also been caused by explosive volcanic eruptions. When the dust from an explosive volcanic eruption enters the atmosphere, wind currents can spread the dust around the globe. This dust can create a shroud over the earth, which can reflect solar radiation back into space, thus cooling the earth. It has been measured that a single explosive eruption may cause a $0.2°C$ to $0.3°C$ cooling in temperature.

The increase in greenhouse gases has been a perpetrator in climate change. The addition of greenhouse gases in the atmosphere is causing a general increase in the earth's temperature. The burning of fossil fuels has been implicated in the recent increases in the earth's temperature. Greenhouse gases, however, can also be increased during periods of heightened out gassing of the lithosphere. Global warming, due to the increase of greenhouse gasses in the atmosphere, causes melting of ice caps, sea level rise, increased storm activity, and changes in ocean temperature patterns (as seen with El Niño and La Niña).

Plate tectonics
Plate tectonics are caused by convection currents in the asthenosphere of the Earth. The asthenosphere is an amorphous solid, which means that it is able to flow. Heat from deep within the earth causes the asthenosphere to warm. The warm material rises to the surface. The material then forms volcanic eruptions, adding to the earth's crust. This causes a divergent plate boundary to form, pushing the lithospheric plates apart. The asthenosphere cools as it rises. In the cooler areas of the asthenosphere, material begins to sink, causing a convergent plate boundary.

Skill 15.2   Identify and explain Earth processes and cycles and cite examples in real-life situations, including conceptual questions on rock cycles, volcanism, and plate tectonics.

Data obtained from many sources led scientists to develop the theory of plate tectonics. This theory is the most current model that explains not only the movement of the continents, but also the changes in the earth's crust caused by internal forces.

Plates are rigid blocks of earth's crust and upper mantle. These solid blocks make up the lithosphere. The earth's lithosphere is broken into nine large sections and several small ones. These moving slabs are called plates. The major plates are named after the continents they are "transporting."

The plates float on and move with a layer of hot, plastic-like rock in the upper mantle. Geologists believe that the heat currents circulating within the mantle cause this plastic zone of rock to slowly flow, carrying along the overlying crustal plates.

Movement of these crustal plates creates areas where the plates diverge as well as areas where the plates converge. A major area of divergence is located in the Mid-Atlantic. Currents of hot mantle rock rise and separate at this point of divergence creating new oceanic crust at the rate of 2 to 10 centimeters per year. Convergence is when the oceanic crust collides with either another oceanic plate or a continental plate. The oceanic crust sinks forming an enormous trench and generating volcanic activity. Convergence also includes continent to continent plate collisions. When two plates slide past one another a transform fault is created.

Orogeny is the term given to natural mountain building.

A mountain is terrain that has been raised high above the surrounding landscape by volcanic action, or some form of tectonic plate collisions. The plate collisions could be intercontinental or ocean floor collisions with a continental crust (subduction). The physical composition of mountains would include igneous, metamorphic, or sedimentary rocks; some may have rock layers that are tilted or distorted by plate collision forces.

There are many different types of mountains. The physical attributes of a mountain range depends upon the angle at which plate movement thrust layers of rock to the surface. Many mountains (Adirondacks, Southern Rockies) were formed along high angle faults.

Folded mountains (Alps, Himalayas) are produced by the folding of rock layers during their formation. The Himalayas are the highest mountains in the world and contain Mount Everest, which rises almost 9 km above sea level. The Himalayas were formed when India collided with Asia. The movement that created this collision is still in process at the rate of a few centimeters per year.

Fault-block mountains (Utah, Arizona, and New Mexico) are created when plate movement produces tension forces instead of compression forces. The area under tension produces normal faults and rock along these faults is displaced upward.

Dome mountains are formed as magma tries to push up through the crust but fails to break the surface. Dome mountains resemble a huge blister on the earth's surface.

Upwarped mountains (Black Hills of South Dakota) are created in association with a broad arching of the crust. They can also be formed by rock thrust upward along high angle faults.

Faults are categorized on the basis of the relative movement between the blocks on both sides of the fault plane. The movement can be horizontal, vertical or oblique.

A dip-slip fault occurs when the movement of the plates is vertical and opposite. The displacement is in the direction of the inclination, or dip, of the fault. Dip-slip faults are classified as normal faults when the rock above the fault plane moves down relative to the rock below.

Reverse faults are created when the rock above the fault plane moves up relative to the rock below. Reverse faults having a very low angle to the horizontal are also referred to as thrust faults.

Faults in which the dominant displacement is horizontal movement along the trend or strike (length) of the fault are called **strike-slip faults**. When a large strike-slip fault is associated with plate boundaries it is called a **transform fault**. The San Andreas Fault in California is a well-known transform fault.

Faults that have both vertical and horizontal movement are called **oblique-slip faults.**

Volcanism is the term given to the movement of magma through the crust and its emergence as lava onto the earth's surface. Volcanic mountains are built up by successive deposits of volcanic materials.

An active volcano is one that is presently erupting or building to an eruption. A dormant volcano is one that is between eruptions but still shows signs of internal activity that might lead to an eruption in the future. An extinct volcano is said to be no longer capable of erupting. Most of the world's active volcanoes are found along the rim of the Pacific Ocean, which is also a major earthquake zone. This curving belt of active faults and volcanoes is often called the Ring of Fire. The world's best known volcanic mountains include Mount Etna in Italy and Mount Kilimanjaro in Africa. The Hawaiian Islands are actually the tops of a chain of volcanic mountains that rise from the ocean floor.

There are three types of volcanic mountains: shield volcanoes, cinder cones, and composite volcanoes.

Shield Volcanoes are associated with quiet eruptions. Lava emerges from the vent or opening in the crater and flows freely out over the earth's surface until it cools and hardens into a layer of igneous rock. A repeated lava flow builds this type of volcano into the largest volcanic mountain. Mauna Loa found in Hawaii, is the largest volcano on earth.

Cinder Cone Volcanoes are associated with explosive eruptions as lava is hurled high into the air in a spray of droplets of various sizes. These droplets cool and harden into cinders and particles of ash before falling to the ground. The ash and cinder pile up around the vent to form a steep, cone-shaped hill called the cinder cone. Cinder cone volcanoes are relatively small but may form quite rapidly.

Composite Volcanoes are described as being built by both lava flows and layers of ash and cinders. Mount Fuji in Japan, Mount St. Helens in Washington, USA, and Mount Vesuvius in Italy are all famous composite volcanoes.

When lava cools, **igneous rock** is formed. This formation can occur either above ground or below ground.

Intrusive rock includes any igneous rock that was formed below the earth's surface. Batholiths are the largest structures of intrusive type rock and are composed of near granite materials; they are at the core of the Sierra Nevada Mountains.

Extrusive rock includes any igneous rock that was formed at the earth's surface.

Dikes are old lava tubes formed when magma entered a vertical fracture and hardened. Sometimes magma squeezes between two rock layers and hardens into a thin horizontal sheet called a **sill**. A **laccolith** is formed in much the same way as a sill, but the magma that creates a laccolith is very thick and does not flow easily. It pools and forces the overlying strata up, creating an obvious surface dome.

A **caldera** is normally formed by the collapse of the top of a volcano. This collapse can be caused by a massive explosion that destroys the cone and empties most, if not all, of the magma chamber below the volcano. The cone collapses into the empty magma chamber forming a caldera.

An inactive volcano may have magma solidified in its pipe. This structure, called a volcanic neck, is resistant to erosion and today may be the only visible evidence of the past presence of an active volcano.

Skill 15.3   Analyze the transfer of energy within and among Earth's land, water, and atmospheric systems, including the identification of energy sources of volcanoes, hurricanes, thunderstorms, and tornadoes.

The sun is considered the nearest star to earth that produces solar energy. By the process of nuclear fusion, hydrogen gas is converted to helium gas. Energy flows out of the core to the surface, then radiation escapes into space. Solar radiation is energy traveling from the sun that radiates into space. Solar flares produce excited protons and electrons that shoot outward from the chromosphere at great speeds reaching earth. These particles disturb radio reception and also affect the magnetic field on earth.

The movement of ocean water is caused by the wind, the sun's heat energy, the earth's rotation, the moon's gravitational pull on earth, and by underwater earthquakes. Most ocean waves are caused by the impact of winds. Wind blowing over the surface of the ocean transfers energy (friction) to the water and causes waves to form. Waves are also formed by seismic activity on the ocean floor. A wave formed by an earthquake is called a seismic sea wave. These powerful waves can be very destructive, with wave heights increasing to 30 m or more near the shore.

A **thunderstorm** is a brief, local storm produced by the rapid upward movement of warm, moist air within a cumulo-nimbus cloud. Thunderstorms always produce lightning and thunder, accompanied by strong wind gusts and heavy rain or hail.

A severe storm with swirling winds that may reach speeds of hundreds of km per hour is called a **tornado**. Such a storm is also referred to as a "twister". The sky is covered by large cumulo-nimbus clouds and violent thunderstorms; a funnel-shaped swirling cloud may extend downward from a cumulonimbus cloud and reach the ground. Tornadoes are narrow storms that leave a narrow path of destruction on the ground.

A swirling, funnel-shaped cloud that **extends** downward and touches a body of water is called a **waterspout.**

Hurricanes are storms that develop when warm, moist air carried by trade winds rotates around a low-pressure "eye". A large, rotating, low-pressure system accompanied by heavy precipitation and strong winds is called a tropical cyclone or is better known as a hurricane. In the Pacific region, a hurricane is called a typhoon.

Storms that occur only in the winter are known as blizzards or ice storms. A **blizzard** is a storm with strong winds, blowing snow and frigid temperatures. An **ice storm** consists of falling rain that freezes when it strikes the ground, covering everything with a layer of ice.

Wind, rain, hail, snow, lightning, tornadoes and hurricanes are all results of energy transformations brought about by solar energy on the Earth. The average total solar incoming radiation (or insolation) is estimated to be about 1350 watts per square meter at the equator at midday, a figure known as the solar constant (although this amount varies a little each year, as a result of solar flares, prominences and the sunspot cycle). Some 34% of this is immediately reflected by the planetary albedo, as a result of clouds, snowfields, and even reflected light from water, rock or vegetation.

More energy is received in the tropics than is re-radiated while more energy is radiated at the poles than is received; therefore, climatic homeostasis is only maintained by a transfer of energy from the tropics to the poles. This transfer of energy drives the winds and the ocean currents. All meteorological processes involve transformation of energy from a concentrated form (such as sunlight) into a less concentrated form (such as infrared radiation). Energy may be temporarily locally stored during this process, and the sudden release of such stored sources is responsible for the dramatic processes mentioned above. For example, the kinetic energy of a snow-avalanche or hurricane is due to the sudden release of energy previously captured from solar radiations.

Continental drift, mountain ranges, volcanoes, and earthquakes are phenomena that are a result of energy transformations in the Earth's crust. The rate of energy transformations inside the Earth is about 37×10^{12} J/s (37 terawatts). This is only about 0.1% of the amount of energy Earth receives from Sun in the form of sunlight and radiates back into space in the form of IR blackbody radiation (~4×10^{16} J/s = 40 petawatts).

Scientists have estimated that about 24 terawatts (65%) of this rate of energy transformation is due to radioactive decay (principally of potassium 40, thorium 232 and uranium 238), and the remaining 13 terawatts is from the continuous gravitational sorting of the core and mantle of the earth, energies left over from the formation of the Earth. It has been estimated that the current radioactive energy of the planet represents less than 1% of that which was available at the time the planet was formed.

As a result, geological forces of continental accretion, subduction and sea floor spreading, account for 90% of the Earth's energy. The remaining 10% of geological tectonic energy comes through hotspots produced by mantle plumes, resulting in shield volcanoes like Hawaii, geyser activity like Yellowstone or flood basalts like Iceland.

Tectonic processes are driven by heat from the Earth's interior. The uplifting of metamorphosed weathered rocks creates mountain ranges. The potential energy represented by the mountain range's weight and height thus represents heat from the core of the Earth. Since this is potential energy, it may be suddenly released in landslides or tsunamis. Similarly, the energy release which drives an earthquake represents stresses in rocks that are mechanical potential energy that was stored from tectonic processes. In summary, an earthquake thus ultimately represents kinetic energy which is being released from elastic potential energy in rocks, which in turn has been stored from heat energy released by radioactive decay and gravitational collapse in the Earth's interior.

The energy that is responsible for erosion and deposition is a result of the interaction of solar energy and gravity. An estimated 23% of the total insolation is used to drive the water cycle. When water vapor condenses to fall as rain, it dissolves small amounts of carbon dioxide, making a weak hydrochloric acid. This acid acts upon the metallic silicates that form most rocks, producing chemical weathering, removing the metals, and leading to the production of rocks and sand which get carried by wind and water downslope through gravity to be deposited at the edge of continents in the sea. Physical weathering of rocks is produced by the expansion of ice crystals in the joint planes of rocks. The geologic cycle is continued when these eroded rocks are later uplifted into mountains.

There is no absolute measure of energy. Rather, energy is measured in terms of the transition of a system from one state into another. The methods for the measurement of energy often deploy methods for the measurement of still more fundamental concepts of science such as heat, radiation, temperature, electric charge, electric current, and work.

Skill 15.4 Explain the functions and applications of the instruments and technologies used to study the earth sciences, including seismographs, barometers, and satellite systems.

Satellites have improved our ability to communicate and transmit radio and television signals. Navigational abilities have been greatly improved through the use of satellite signals. Sonar uses sound waves to locate objects and is especially useful underwater. The sound waves bounce off the object and are used to assist in location. **Seismographs** record vibrations in the earth and allow us to measure earthquake activity. Common instruments for that forecasting weather include the **aneroid barometer** and the **mercury barometer,** which both measure air pressure. In the aneroid barometer, the air exerts varying pressures on a metal diaphragm that will then read air pressure. The mercury barometer operates when atmospheric pressure pushes on a pool of mercury in a glass tube. The higher the pressure, the higher up the tube mercury will rise. Relative humidity is measured by two kinds of additional weather instruments, the psychrometer and the hair gygrometer.

COMPETENCY 16.0 UNDERSTAND AND APPLY KNOWLEDGE OF OBJECTS IN THE UNIVERSE AND THEIR DYNAMIC INTERACTIONS.

Skill 16.1 Describe and explain the relative and apparent motions of the sun, the moon, stars, and planets in the sky.

Astronomers use groups or patterns of stars called **constellations** as reference points to locate other stars in the sky. Familiar constellations include Ursa Major (also known as the big bear) and Ursa Minor (known as the little bear). Within the Ursa Major, the smaller constellation, The Big Dipper is found. Within the Ursa Minor, the smaller constellation, The Little Dipper is found.

Different constellations appear as the earth continues its revolution around the sun with the seasonable changes.

Magnitude stars are 21 of the brightest stars that can be seen from earth, these are the first stars noticed at night. In the Northern Hemisphere there are 15 commonly observed first magnitude stars.

Doppler Effect: the apparent change in frequency of light or sound that occurs when the source of the wave is moving relative to the observer. The Doppler effect can be used to measure the motion of an object both on Earth and in space.

The motion of the stars can be determined by observing the Doppler effect associated with their movements. As a star moves away from us it has a **Red Shift** and its emitted light wavelength is longer. As a star moves toward us, it has a **Blue Shift** and its emitted light wavelength is shorter. *Red and Blue shift doesn't refer to the color of the star.* It refers to the movement of the λ along the spectrum. By comparing the λ of a star at different times, we are able to determine its shift. We find the λ shift by examining the object's spectral properties. Doppler can also detect rotation. We get differing wavelengths because one side is closer and the other side is farther away.

Skill 16.2 Recognize properties of objects (e.g., comets, asteroids) within the solar system and their dynamic interactions.

Astronomers believe that the rocky fragments that may have been the remains of the birth of the solar system that never formed into a planet. **Asteroids** are found in the region between Mars and Jupiter.

Comets are masses of frozen gases, cosmic dust, and small rocky particles. Astronomers think that most comets originate in a dense comet cloud beyond Pluto. Comet consists of a nucleus, a coma, and a tail. A comet's tail always points away from the sun. The most famous comet, **Halley's Comet,** is named after the person whom first discovered it in 240 B.C. It returns to the skies near earth every 75 to 76 years.

Meteoroids are composed of particles of rock and metal of various sizes. When a meteoroid travels through the earth's atmosphere, friction causes its surface to heat up and it begins to burn. The burning meteoroid falling through the earth's atmosphere is now called a **meteor** or also known as a "shooting star." **Meteorites** are meteors that strike the earth's surface. A physical example of the impact of the meteorite on the earth's surface can be seen in Arizona, The Barringer Crater is a huge Meteor Crater. There many other such meteor craters found throughout the world.

Skill 16.3 Recognize the types, properties, and dynamics of objects external to the solar system (e.g., black holes, supernovas, galaxies).

A vast collection of stars is defined as **galaxies**. Galaxies are classified as irregular, elliptical, and spiral. An irregular galaxy has no real structured appearance; most are in their early stages of life. An elliptical galaxy is smooth ellipses, containing little dust and gas, but composed of millions or trillion stars. Spiral galaxies are disk-shaped and have extending arms that rotate around its dense center. Earth's galaxy is found in the Milky Way and it is a spiral galaxy.

A **pulsar** is defined as a variable radio source that emits signals in very short, regular bursts; believed to be a rotating neutron star.

A **quasar** is defined as an object that photographs like a star but has an extremely large redshift and a variable energy output; believed to be the active core of a very distant galaxy.

Black holes are defined as an object that has collapsed to such a degree that light can not escape from its surface; light is trapped by the intense gravitational field.

A **supernova** is the result of the massive explosion of an upper main sequence Supergiant star caused by the detonation of carbon within the star. A supernova releases more energy than Earth's sun will produce in its entire life cycle. The luminosity of a supernova is as bright as 500 million Suns. For example: Chinese astronomers in 1054 recorded the sudden appearance of a new star in what is now known as the Taurus Constellation. Bright enough to be seen during daytime for over a month, it remained visible for over 2 years.

TEACHER CERTIFICATION STUDY GUIDE

COMPETENCY 17.0 UNDERSTAND AND APPLY KNOWLEDGE OF THE ORIGINS OF AND CHANGES IN THE UNIVERSE.

Skill 17.1 Identify scientific theories dealing with the origin of the universe (e.g., big bang).

Two main theories to explain the origins of the universe include **The Big Bang Theory and The Steady-State Theory.**

The Big Bang Theory has been widely accepted by many astronomers. It states that the universe originated from a magnificent explosion spreading mass, matter and energy into space. The galaxies formed from this material as it cooled during the next half-billion years.

The Steady-State Theory is the least accepted theory. It states that the universe is a continuously being renewed. Galaxies move outward and new galaxies replace the older galaxies. Astronomers have not found any evidence to prove this theory.

The future of the universe is hypothesized with the Oscillating Universe Hypothesis. It states that the universe will oscillate or expand and contract. Galaxies will move away from one another and will in time slow down and stop. Then a gradual moving toward each other will again activate the explosion or The Big Bang theory.

Skill 17.2 Analyze evidence relating to the origin and physical evolution of the universe (e.g., microwave background radiation, expansion).

Cosmic microwave background radiation (CMBR) is the oldest light we can see. It is a snapshot of how the universe looked in its early beginnings. First discovered in 1964, CMBR is composed of photons which we can see because of the atoms that formed when the universe cooled to 3000 K. Prior to that, after the Big Bang, the universe was so hot that the photons were scattered all over the universe, making the universe opaque. The atoms caused the photons to scatter less and the universe to become transparent to radiation. Since cooling to 3000K, the universe has continued to expand and cool.

COBE, launched in 1989, was the first mission to explore slight fluctuations in the background. WMAP, launched in 2001, took a clearer picture of the universe, providing evidence to support the Big Bang Theory and add details to the early conditions of the universe. Based upon this more recent data, scientists believe the universe is about 13.7 billion years old and that there was a period of rapid expansion right after the Big Bang. They have also learned that there were early variations in the density of matter resulting in the formation of the galaxies, the geometry of the universe is flat, and the universe will continue to expand forever.

Skill 17.3   Compare the physical and chemical processes involved in the life cycles of objects within galaxies.

Scientists believe that **stars** form when compression waves traveling through clouds of gas create knots of gas in the clouds. The force of gravity within these denser areas then attracts gas particles. As the knot grows, the force increases and attracts more gas particles, eventually forming a large sphere of compressed gas with internal temperatures reaching a few million degrees C. At these temperatures, the gases in the knot become so hot that nuclear fusion of hydrogen to form helium takes place, creating large amounts of nuclear energy and forming a new star. Pressure from the radiation of these new stars causes more knots to form in the gas cloud, initiating the process of creating more stars.

Scientists theorize that **planets** form from gas and dust surrounding young stars. As the density of the forming star increases, this gas and dust slowly condenses into a spinning disk. The denser areas of the disk develop a gravitational force which attracts more dust and gas as the disk orbits the star. Over millions of years, these dense areas consolidate and grow, forming planets. In the case of the Sun, the larger icy fragments surrounding it attracted more gas and dust forming the more massive planets such as Jupiter and Saturn. These larger planets developed gravitational forces great enough to attract hydrogen and helium atoms, turning them into gas giants. The smaller planets, such as Earth, could not attract these atoms and became mainly rocky.

It is believed that **black holes** form as stars evolve. As the nuclear fuels are used up in the core of a star, the pressure associated with the production of these fuels no longer exists to resist contraction of the core. Two new types of pressure, electron and neutron, arise. However, if the star is more than about five times as massive as the Sun, neither pressure will prevent the star from collapsing into a black hole.

When the universe was forming, most of the material became concentrated in the planets and moons. There were however many small, rocky objects called **planetesimals** that also formed from the gas and dust. These planetesimals include **comets** and **asteroids**. A large cloud of comets, known as the Oort cloud, exists beyond Pluto. A change in the gravitational pull of our galaxy may disturb the orbit of a comet causing it to fall toward the Sun. The ice in the comet turns into vapor, releasing dust from the body. Gas and dust then form the tail of the comet. In the early life of the solar system, some of the planetesimals came together more toward the center of the solar system. The gravitational pull of Jupiter prevented these planetesimals from developing into full planets. They broke up into thousands of minor planets, known as asteroids.

It is believed that **black holes** form as stars evolve. As the nuclear fuels are used up in the core of a star, the pressure associated with the production of these fuels no longer exists to resist contraction of the core. Two new types of pressure, electron and neutron, arise. However, if the star is more than about five times as massive as the Sun, neither pressure will prevent the star from collapsing into a black hole.

When the universe was forming, most of the material became concentrated in the planets and moons. There were however many small, rocky objects called **planetesimals** that also formed from the gas and dust. These planetesimals include **comets** and **asteroids**. A large cloud of comets, known as the Oort cloud, exists beyond Pluto. A change in the gravitational pull of our galaxy may disturb the orbit of a comet causing it to fall toward the Sun. The ice in the comet turns into vapor, releasing dust from the body. Gas and dust then form the tail of the comet. In the early life of the solar system, some of the planetesimals came together more toward the center of the solar system. The gravitational pull of Jupiter prevented these planetesimals from developing into full planets. They broke up into thousands of minor planets, known as asteroids.

Skill 17.4   Explain the functions and applications of the instruments, technologies, and tools used in the study of the space sciences, including the relative advantages and disadvantages of earth-based versus space-based instruments and optical versus non-optical instruments.

Types of **telescopes** used in the study of the space sciences include optical, radio, infrared, ultraviolet, x-ray, and gamma-ray. Optical telescopes work by collecting and magnifying visible light that is given off by stars or reflected from the surfaces of the planets. However, stars also give off other types of electromagnetic radiation, including radio waves, microwaves, infrared light, ultraviolet light, X rays, and gamma rays. Therefore, specific types of non-optical instruments have been developed to collect information about the universe through these other types of electromagnetic waves.

Many of the telescopes used by astronomers are earth-based, located in observatories around the world. However, only radio waves, visible light, and some infrared radiation can penetrate our atmosphere to reach the earth's surface. Therefore, scientists have launched telescopes into space, where the instruments can collect other types of electromagnetic waves. Space probes are also able to gather information from distant parts of the solar system.

In addition to telescopes, scientists construct mathematical models and computer simulations to form a scientific account of events in the universe. These models and simulations are built using evidence from many sources, including the information gathered through telescopes and space probes.

SUBAREA V. THE EARTH AND ATMOSPHERE

COMPETENCY 18.0 UNDERSTAND AND APPLY KNOWLEDGE OF THE GEOLOGIC PROCESSES AND STRUCTURE OF EARTH.

Skill 18.1 Relate the dynamic processes that occur in Earth's interior to their causes (e.g., thermal convection) and effects (e.g., earthquakes, volcanic eruptions).

As the Earth's core produces intense heat, the heat rises through the layers. The heating, rising, cooling, and sinking of the molten, semi-plastic material in the asthenosphere, sets up **convection cells** that are constantly in motion.

When the rising material is "churned" upward by the motion of the convection cells, it hits the bottom of the lithospheric plates, where most of the material moves sideways, cools, and then sinks downward toward the core. However, some of it squeezes upward through cracks in the lithosphere. This material is referred to as Magma.

How far upward the magma moves is generally dependent on the temperature and density of the magma, as well as the size of the crack.

The laws of physics and chemistry tell us that pressure increases when a substance is forced into a confined space. As the magma moves upward, the increasing pressure can cause "ripping" effect that enlarges the crack, pushing the lithospheric materials apart.

Magma is produced at Hot Spots, Spreading Centers, and Subduction Zones and varies in composition according to where it is produced.

Hot Spot and Spreading Center: The magma is produced in the asthenosphere and comes up mantle plume. This magma is pure magma, dense and silica poor. However overall, the material is very runny and fluid.

Subduction Zone: The magma is impure because ocean sediment is sucked down and mixed with the magma. The silica rich, water saturated magma is thicker because of the addition of ocean sediment.

Ocean/Ocean Boundaries: These plates are forced together by the spreading of the ocean floor. Tremendous frictional forces are created as the plates collide and some of the oceanic material builds upward, while other oceanic material bends downward.

The leading edges of the boundaries meet around 700 km downward. The forces involved push some material upward through the lithosphere to become a volcano. The built up materials may eventually break the ocean surface to become volcanic islands. This effect is actually so widespread that the islands form groupings of volcanic islands called Volcanic Arcs.

Example: The Philippines, Japan, and Guam.

The Volcanism is of the explosive type and the quick release of fantastic strain and pressure causes devastating **Deep Focus Earthquakes**. Fantastic strain and pressures cause the devastating deep focus earthquakes.

Example: Volcanism: The 1993 Mt. Pinatubo eruption in the Philippines.
Example: Earthquake: The Easter, 1964, Alaskan earthquake measured 9.2 on Richter scale.

Hot Spot: an offshoot of a convection cell that burns through the lithospheric material in the middle, not at the edges, of a plate.

Aseismic Ridge: a series of volcanoes (underwater or above the surface) formed by the movement of the plates over a hot spot. They usually form a dogleg pattern, showing the change of direction in the plates' movement.

Example: The Hawaiian Islands.

Skill 18.2 Identify the basic principles of plate tectonic theory, supporting evidence for the theory, and the mechanisms of plate dynamics.

Data obtained from many sources led scientists to develop the theory of plate tectonics. This theory is the most current model that explains not only the movement of the continents, but also the changes in the earth's crust caused by internal forces.

Plates are rigid blocks of earth's crust and upper mantle. These solid blocks make up the lithosphere. The earth's lithosphere is broken into nine large sections and several small ones. These moving slabs are called plates. The major plates are named after the continents they are "transporting."

The plates float on and move with a layer of hot, plastic-like rock in the upper mantle. Geologists believe that the heat currents circulating within the mantle cause this plastic zone of rock to slowly flow, carrying along the overlying crustal plates.

Movement of these crustal plates creates areas where the plates diverge as well as areas where the plates converge. A major area of divergence is located in the Mid-Atlantic. Currents of hot mantle rock rise and separate at this point of divergence creating new oceanic crust at the rate of 2 to 10 centimeters per year. Convergence is when the oceanic crust collides with either another oceanic plate or a continental plate. The oceanic crust sinks forming an enormous trench and generating volcanic activity. Convergence also includes continent to continent plate collisions. When two plates slide past one another a transform fault is created.

These movements produce many major features of the earth's surface, such as mountain ranges, volcanoes, and earthquake zones. Most of these features are located at plate boundaries, where the plates interact by spreading apart, pressing together, or sliding past each other. These movements are very slow, averaging only a few centimeters each year.

Boundaries form between spreading plates where the crust is forced apart in a process called rifting. Rifting generally occurs at mid-ocean ridges. Rifting can also take place within a continent, splitting the continent into smaller landmasses that drift away from each other, thereby forming an ocean basin between them. The Red Sea is a product of rifting. As the seafloor spreading takes place, new material is added to the inner edges of the separating plates. In this way the plates grow larger, and the ocean basin widens. This is the process that broke up the super continent Pangaea and created the Atlantic Ocean.

Boundaries between plates that are colliding are zones of intense crustal activity. When a plate of ocean crust collides with a plate of continental crust, the more dense oceanic plate slides under the lighter continental plate and plunges into the mantle. This process is called **subduction**, and the site where it takes place is called a subduction zone. A subduction zone is usually seen on the sea-floor as a deep depression called a trench.

The crustal movement that is identified by plates sliding sideways past each other produces a plate boundary characterized by major faults that are capable of unleashing powerful earth-quakes. The San Andreas Fault forms such a boundary between the Pacific Plate and the North American Plate.

Skill 18.3 Relate the features and landforms of Earth's surface to the processes that shape them (e.g., uplift, erosion).

Plates are rigid blocks of earth's crust and upper mantle. These rigid solid blocks make up the lithosphere. The earth's lithosphere is broken into nine large sections and several small ones. These moving slabs are called plates. The major plates are named after the continents they are "transporting."

The plates float on and move with a layer of hot, plastic-like rock in the upper mantle. Geologists believe that the heat currents circulating within the mantle cause this plastic zone of rock to slowly flow, carrying along the overlying crustal plates.

Movement of these crustal plates creates areas where the plates diverge as well as areas where the plates converge. A major area of divergence is located in the Mid-Atlantic. Currents of hot mantle rock rise and separate at this point of divergence creating new oceanic crust at the rate of 2 to 10 centimeters per year. Convergence is when the oceanic crust collides with either another oceanic plate or a continental plate. The oceanic crust sinks forming an enormous trench known as a subduction zone and generating volcanic activity. Portions of the lithosphere are dragged into the mantle. Then some of this material melts and volcanoes erupt. In time, series of volcanic islands are formed like the Aleutian Islands which are parallel to the trench.

Convergence also includes continent to continent plate collisions. When two plates slide past one another a transform fault is created. The crustal movement which is identified by plates sliding sideways past each other produces a plate boundary characterized by major faults that are capable of unleashing powerful earthquakes. The San Andreas Fault forms such a boundary between the Pacific Plate and the North American Plate.

A mountain is terrain that has been raised high above the surrounding landscape by volcanic action, or some form of tectonic plate collisions. This is known as uplift. The plate collisions could be intercontinental or ocean floor collisions with a continental crust (subduction). The physical composition of mountains would include igneous, metamorphic, or sedimentary rocks; some may have rock layers that are tilted or distorted by plate collision forces.

There are many different types of mountains. The physical attributes of a mountain range depends upon the angle at which plate movement thrust layers of rock to the surface. Many mountains (Adirondacks, Southern Rockies) were formed along high angle faults.

Folded mountains (Alps, Himalayas) are produced by the folding of rock layers during their formation. The Himalayas are the highest mountains in the world and contain Mount Everest which rises almost 9 km above sea level. The Himalayas were formed when India collided with Asia. The movement which created this collision is still in process at the rate of a few centimeters per year.

Fault-block mountains (Utah, Arizona, and New Mexico) are created when plate movement produces tension forces instead of compression forces. The area under tension produces normal faults and rock along these faults is displaced upward.

Dome mountains are formed as magma tries to push up through the crust but fails to break the surface. Dome mountains resemble a huge blister on the earth's surface.

Upwarped mountains (Black Hills of S.D.) are created in association with a broad arching of the crust. They can also be formed by rock thrust upward along high angle faults.

Volcanism is the term given to the movement of magma through the crust and its emergence as lava onto the earth's surface. Volcanic mountains are built up by successive deposits of volcanic materials.

1. Weathering

Weathering: the physical and chemical breakdown and alteration of rocks and minerals at or near the Earth's surface.

Classifying Weathering:

Mechanical (Also called Physical Weathering: where rock is broken into smaller pieces with no change in chemical or mineralogical composition. The resulting material still resembles the original material.
Example: Rock pieces breaking off a boulder.
The pieces still resemble the original material, but on a smaller scale.

Chemical Weathering: where a chemical or mineralogical change occurs in the rock and the resulting material no longer resembles the original material.
Example: Granite (Gneiss/Schist) eventually weathers into separate sand, silt, and clay particles.

Weathering is typically caused by a combination of chemical and mechanical processes.

Factors Influencing Weathering:

Composition: Due to their composition, some rocks weather easier and will show more effects.

Rock Structure: Does it have cracks? Is it fractured? Water and other elements get in the cracks.

Climate: The more water, the more the weathering effect. Additionally, the higher the temperature or the more the temperature varies, the greater the weathering effect.

Topography: This factor determines the amount of surface area exposed to weathering. Smaller rocks are affected more because collectively, they have less mass and more surface area than a boulder.

Vegetation: Important weathering agent. Depending on the type of vegetation, it can either hinder or accelerate the weathering process. Although vegetation may leave less surface area exposed, the vegetation's root structures can produce a biological effect that accelerates the process.

Types of Mechanical Weathering:

Frost Wedging: This occurs when rock gets a crack in it, water collects, and then freezes. Over time, as this cycle repeats itself, the expanding water gradually pushes the rock apart.

Salt Crystal Growth: In a process similar to frost wedging, as the water evaporates, it leaves salt crystals behind. Eventually, these crystals build up an push the rock apart. This is a very small-scale effect and takes considerably longer than frost wedging to affect the rock material.

Abrasion: This is a key factor in mechanical weathering. The motion of the landscape materials produces significant weathering effects, scouring, chipping, or wearing away pieces of material.

Abrasive agents include wind blown sand, water movement, and the materials in landslides bashing into each other.

Biological Activity: This is a two-fold weathering agent.

Plants: Seeds will sometimes land in a crack in a rock and begin to grow in the cracks. The root structure eventually acts as a wedge, pushing the rock apart.

Animal: As animals burrow, the displaced material has an abrasive effect on the surrounding rock. Because of the limited number of burrowing animals, plant activity has a much greater weathering effect.

Pressure Release (Exfoliation): Rock expands when compressive forces are removed, and bits of the rock break off during expansion. This can result in massive rock formations with rounded edges.
Example: Half-Dome in Yosemite National Park.

Thermal Expansion and Contraction: Minerals within a rock will expand or contract due to changes in temperature. Dependent on the minerals in the rock, this expansion and contraction occurs at different rates and to different magnitudes. Essentially, the rock internally tears itself apart. The rock may look solid but when placed under pressure, easily crumbles.

Climate is a key factor in mechanical weathering.

Types of Chemical Weathering:

Oxidation (Rust): Oxygen atoms become incorporated into the formula of a mineral in a rock and the mineral becomes unstable and breaks off in flakes. Example: Iron oxide (FeO_2) changes to iron trioxide (FeO_3) due to the oxygen chemically imparted to the mineral.

Solution: Due to their inherent composition, some minerals found in rocks easily dissolve into solution when exposed to a liquid. Example: Halite (Rock Salt) completely dissolves in water.

Acids: Water and water vapor may combine with other elements and gases to form acids.

Water (H_2O) and carbon dioxide (CO_2) can chemically combine to become Carbonic Acid (H_2CO_3).

Sulfur Dioxide (SO_2) particles can chemically combine with water (H_2O) to form Sulfuric Acid (H_2SO_4).

Biological Activity: Plant roots growing in the cracks of rocks not only cause mechanical wedging, but also secrete acids that cause chemical weathering.

Generally found in combination with solution, *acids* cause the majority of chemical weathering.

Common Effects of Weathering

Quartz: Inert to most effects, quartz stays in the environment as original material.

Feldspar: forms clay minerals.

Ferromags: Also form clay minerals. Example: Gneiss/Schist can erode to clay minerals as the Quarts is leeched out, and blown or transported away.

Carbonates: Highly susceptible to carbonic acids, carbonates and carbonated rocks such as limestone will – dependent on the concentration of acid – will dissolve into solution and/or become riddled with holes.

Differential Weathering:

Differential weathering occurs when rocks on a landscape *weather at different rates*. The effect of this difference can create fantastic landscape features, with some rocks "sticking up," arched, or nested in deep depressions. Examples: Ship Rock, Devil's Tower, and the other mesa and butte formations in Monument Valley.

2. Stream Erosion

The action of the stream causes various erosive effects:

Hydraulic Action: It smoothes rocks and carves potholes.

Potholes: Holes bored into rock material by the swirling (hydraulic) action of a stream.

Abrasion: Sand particles carried by the water acts like sandpaper, cutting and smoothing the landscape material.

Moving water in a stream has two major effects: It carves downwards and sideways. The downward carving is the first and most prominent action caused by a moving stream. It erodes the banks and beds of a stream, gradually changing the shape of the stream and its surrounding landscape. Example: The Grand Canyon.

Base Level: the lowest level to which a stream can erode.

The geographic location to wherever a stream flows is the base level. As a stream approaches the base level due to downward cutting, the action shifts to a forward motion. The higher the velocity, the more efficient the erosion.

Mountain streams have little fining (sorting the material by size) due to their higher velocity, and low land streams are muddy because the velocity is less and erosion occurs on the bed and sides of the stream.

Once a stream is at or close to base level, equilibrium is achieved between deposition and erosion.

The natural tendency for a stream is to flow in a curved path.
When combined with erosional differences caused by the erosion of sides of the stream into meanders or bends, these meanders grow and migrate downstream through the process of erosion/deposition.

Erosion and deposition are controlled by the velocity of the stream.
As the stream approaches base level, more of its energy is in a side-to-side cutting (meanders) than in down-cutting.

Sometimes after extensive erosion, the meanders become so sinuous that they become cut off and are called cut-off meanders or ox bow lakes.

The lateral erosion develops the floodplain. The older the stream, the flatter the landscape and the closer the stream is to base level.

3. Glacial Erosion

Glaciers are nature's bulldozers. As the glacier moves forward due to basal sliding, the underlying topography is severely abraded because of the immense pressure on the rock base due to the weight of the glacier.

The Ice Age began about 2 -3 million years ago. This age saw the advancement and retreat of glacial ice over millions of years. Theories relating to the origin of glacial activity include Plate Tectonics, where it can be demonstrated that some continental masses, now in temperate climates, were at one time blanketed by ice and snow. Another theory involves changes in the earth's orbit around the sun, changes in the angle of the earth's axis, and the wobbling of the earth's axis. Support for the validity of this theory has come from deep ocean research that indicates a correlation between climatic sensitive micro-organisms and the changes in the earth's orbital status.

There are two main types of glaciers: valley glaciers and continental glaciers. Erosion by valley glaciers is characteristic of U-shaped erosion. They produce sharp peaked mountains such as the Matterhorn in Switzerland. Erosion by continental glaciers often rides over mountains in their paths leaving smoothed, rounded mountains and ridges.

The key factor in the glacial erosion process is the meltwater. It causes frost wedging, which initiates the erosional sequences.

As the glacier moves forward on the meltwater, some of the water seeps into cracks and freezes. This frost wedging causes a further widening and weakening of the rock material. The glacier plucks the fragments out of the cracks and pushes them along the base. These fragments act as a scouring pad, and as they increase in size and amount, they increase the abrasion on the topography. The grinding motion due to the glacial weight increases as the mass moves forward, picking up more fragments and causing **striations** in the underlying rock.

Glaciers can move rock fragments varying in size from pebbles to entire boulders. Fine fragments are referred to as **rock flour**. Boulder sized rock units moved to other locations are called **eratics**.

4. Erosional Features

Alpine Valleys

After a glacier melts, it leaves behind its distinct mark on the landscape. One of the most common of these features is the **Alpine Glacier Valley**. These are usually easy to recognize. A characteristic of glacial landscaping is the U-shaped valley. These usually are situated in previously carved areas caused by streams.

However, the glacier's movement straightens out the curves normally associated with streams and also carves the sides of valley, neatly removing or dramatically shortening any ridges that extend in its path. **Truncated Spurs** – the lowest part of the intruding ridges - are the result of this movement.

Glaciers can also form tributaries similar to streams. However, depending on the size of the glacier, the heavier ice masses can carve downward more rapidly and create much deeper valleys.

Once these glaciers recede, the tributaries remain behind, perched high above the main valley as hanging valleys.

Another typical feature of glaciation are **rock basin lakes**. These are water filled depressions carved in the bedrock by the abrasive action of the glacier's movement. These lakes fill after the glacier melts.

Cirques, Horns and Aretes

Cirque: a steep-sided, rounded hollow at the head of a glacial valley. The cirque is one of the most prominent features associated with Alpine style glaciation.

Horn: the sharp peak formed by the erosional action of cirques. The mountain is cut back on several sides, forming the prominence.

Arete: the sharp ridges that separate the valleys carved by glaciers.

Continental Erosional Features

Till: the fragments that are abraded from the bedrock and deposited along the path of the glacier. The till are usually angular shaped and unsorted.

Moraine: a body of till carried along or deposited by a glacier. This loose material falls from the side of the glacier and accumulates along the edges of the ice path in the valleys.

Lateral Moraine: piles of till along the sides of a glacier.

Medial Moraine: a single long ridge of till carried down the central path of a glacier.

End Moraine: a ridge of till that piles up at the edge of the ice sheet.

Terminal Moraine: a type of end moraine that marks the furthest advance of the glacier.

Recessional Moraine: a type of end moraine that marks the temporary terminus of receding glaciers.

Ground Moraine: an extensive, thin layer of till deposited that creates the rolling topography of areas once affected by glaciation.

Drumlin: a streamlined hill of till shaped like the bowl of an inverted spoon. The long axis of the drumlin is parallel to the direction of the ice movement.

Meltwater Erosional Features

A significant amount of meltwater runs over, beneath, and away from the ice sheet in the zone of wastage. The material deposited by this meltwater is known as **Outwash**.

The materials carried by the outwash behave similar to those moved by a stream; the material is sorted and layered, differentiating it from the unsorted till.

An **Esker**: a long, sinuous ridge comprised of outwash deposited sediment. Can be up to 10 meters in height. The material is deposited in tunnels beneath the ice sheet and left behind as the ice melts.

As the outwash moves sediment alongside and in the path of a receding glacier, blocks of ice can be buried beneath the sediment. After years of erosion these blocks are uncovered and melt, leaving a shallow depression behind. When these depressions fill, they are known as **Kettles**, and become scenic lakes (Kettle Lakes).

The outwash carries a large amount of rock flour and when this eventually settles out and dries, it can be picked up and carried long distances by the wind. This fine-grained dust is called **Loess**.

Sometimes a lake forms between retreating glaciers and end moraines. Clay and silt settle on the bottom in two distinct layers. The light colored layer (silt) represents coarser sediment deposited by glacier melting during the warmer part of the year. The darker colored layer (clay) represents the more slowly settling material that sinks after the lake surface freezes. Taken together, the two-color pattern is called a **Varve**, and indicates the total deposition of material in a one-year span. Like tree-rings, varves can be used as a relative dating technique to roughly estimate the amount of time that the lake has existed.

Plains are vast flat areas that make up about one half the total landscape of the United States. The Atlantic Coastal Plain is the exposed margin of the continental shelf that has emerged from the sea during the past several million years. It is a lowland that rests upon soft, loose sediments such as sand, silt, and clay. Low hills and valleys occur only when the plain has been slightly uplifted. Wide river valleys have cut through the plan producing low hills between river systems. Much of the area is covered by swamps and lakes.

The Great Plains is a young landscape underlain by sediments that have been eroded from the uplifted Rocky Mountains. Streams have deposited these materials in nearly horizontal layers during the last few million years. The plains elevation ranges from 1500 meters at its western boundary to 350 meters at its eastern boundary. Therefore, rivers that cross the plain all flow east. Most of the landforms present in this area are related to the drainage patterns developed by river systems. Many complex drainage patterns have been etched into the loose, soft sediments of the plain.

Stream patterns provide clues to rock resistance in an area. A rock of uniform resistance will produce a dendritic stream pattern. A dendritic pattern resembles a tree with branches, so this is a pattern of creeks flowing into streams, which flow into tributary rivers and eventually into a large river. Radial patterns, streams flowing outwardly from a central location, form on the slopes of volcanoes. Trellis patterns develop in alternating resistant and weak layers of tilted sedimentary rock. And rectangular patterns are seen when the rock is jointed or fractured in a rectangular pattern.

Skill 18.4   Use a geologic column or block diagram to interpret the geologic history of a particular area.

Geologists interpret the order of physical events or the geologic history of an area by observing and creating standardized diagrams of rocks. Because, in most areas, rock formations are covered by vegetation or development, or are underground, geologists must look in certain places to find rock formations that are uncovered and show deeper rocks. These include roadcuts and areas where rock faces have been exposed through weathering. An alternative is drilling into the rock and bringing a core to the surface to be examined.

From these sources of information, a **stratigraphic column** is created:

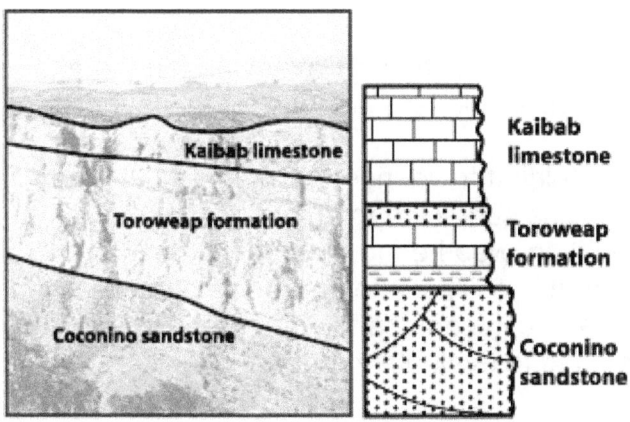

In the diagram above, rocks exposed in the Grand Canyon are translated into a stratigraphic column. This is a fairly simple column, as the most recent rocks are at the top, with the older rocks on the bottom. Because there are a lot of rocks exposed in this area of the western United States, these rock formations have been well-described and are easily recognized by geologists in the field.

The stratigraphic column contains various types of information. In this case, the rock formations are actually named. This allows this rock face to be compared to other exposed rocks distant from this site, which provides a large-scale understanding of the geology of the region. In some cases, such as a stratigraphic column for a core sample in a new area, the rocks would simply be described rather than named. In that case, they would be given names based on the type of rock, such as sandstone, shale, slate, limestone, granite, clay, etc.

Certain universal symbols are used to draw different kinds of rock layers, based on structural characteristics. For example, sandstones are drawn using dots, and limestones are drawn as bricks. From the diagram above, you can see that the Coconino sandstone has also been drawn with additional lines through it, to indicate fracturing. This can be important to geologists looking for water, oil, or natural gas deposits. A thin layer of shale is shown between the Coconino and Toroweap formations as a series of dashed lines.

In other formations, a silty shale or interbedded silts and shales might be shown as dashed lines mixed with dots. Salts may be shown using crosses, and basalts may be shown as black areas, sometimes with vertical lines if they are columnar basalts. Metamorphic rocks may be shown with wavy lines indicating folds; granite rocks as various solid colors with crosses or dots in them. Although there are certain conventions that are typically followed, each area will have its own specific formations and details that will be noted in the legend of a stratigraphic column. Colors are frequently used to indicate the actual color of the rock. Some examples of stratigraphic columns can be found in the links below:

Iowa: http://www.igwa.org/iowacol.asp
Michigan: http://www.deq.state.mi.us/documents/deq-gsd-info-geology-Stratigraphic.pdf

These columns also show the geologic era in which each layer was formed and the thickness of the layer, as well as shorthand letter notations for the layers.

Not all rocks appear from bottom to top, oldest to youngest. Originally they may be deposited this way, but then they can be subjected to a variety of geologic processes, including tilting, folding, uplifting, weathering, fracturing, and upwelling of magma within existing formations. These features can often be seen in roadcuts or other exposed surfaces and can confuse the interpretation of geologic events. Certain features are used to evaluate the order in which events occurred:

- **Rock type/appearance:** First, one should simply look carefully at each layer of rock and evaluate the type of rock, grain size, minerals, color, weathering, layering, and any other identifying characteristics that may be present. These can be compared to stratigraphic records for the area to identify the layer and its age.

- **Layering:** Geologists know that originally, the oldest layers of rocks were laid down on the bottom. This is true whether the rocks were formed through sedimentary processes (e.g., successive layers formed by weathering and erosion) or igneous processes (e.g., layers of lava laid down on top of one another). Due to gravity, these layers are normally laid down in a horizontal manner – i.e., the surface of the rock layers was originally horizontal.

- **Tilting:** Later, tilting of the rocks may have occurred due to processes related to plate tectonics. Generally, it is relatively rare to find rock layers that are tilted more than a 90° angle. Therefore, one can normally assume that the layer on the bottom is still the oldest, even if the layers are tilted.

- **Dating:** Certain types of rock layers can be dated through a number of means. Fossils in the rocks can be compared to known ages of fossils to determine the period in which those organisms lived, to date the rocks. Certain fossils represent organisms that lived only for a short period of time, and these are especially helpful in pinning down the timeframe in which a rock layer was deposited. Radioisotope dating, using a variety of different isotopes depending on the age of the rock layer being dated, can also be used to establish the relative ages of rock formations.

- **Intrusions:** Once layers of rock have been laid down, magma can intrude into the layer, either in large pockets known as batholiths, along fractures or in straight lines known as dikes. The age of these features is interpreted relative to the layers around them. If they cross other layers, then the intrusion is newer than those layers. The image below shows dark vertical dikes crossing nearly horizontal layers of lighter-colored rock in Alaska. Vertical dikes are sometimes called pipes.

- **Weathering:** Weathering can often differentially erode softer layers of rock without eroding harder layers. For example, sedimentary rocks may be eroded before harder igneous rocks. After a long period of weathering, this may leave features such as pipes, dikes, and batholiths exposed while the surrounding rock has completely weathered away. Half-Dome in Yosemite, CA (shown below) is a good example of a batholith that has been exposed by weathering; its round shape indicates its previous history as a magma intrusion.

Subsequently, additional sedimentary or igneous layers may be laid down around the exposed rock. Unfortunately, this can look similar to the case in which a body of magma intrudes into existing layers. Sometimes the margins of the intrusive layer can be examined to evaluate whether they were ever exposed to the elements (which would indicate that the layers surrounding them are newer) or to see if there is evidence of chemical interaction and/or melting of the surrounding rock (which would indicate that the layers surrounding them are pre-existing). In very complicated situations, the various rock layers may need to be compared among multiple locations to be properly interpreted.

Skill 18.5 Demonstrate and explain strategies that are used to identify and classify rocks and minerals.

Sandstone is a particularly good source of information about past environments because of its composition and structure. Composed of sand and other materials, Sandstone is classified by both composition and appearance. There are four major types of sandstone, and these are further classified by the degree of weathering and the presence or absence of ferromags and feldspars.

Types:

1. Quartz Sandstone: Almost pure quartz, very mature, usually indicates stable tectonics.
2. Arkose: >25% Feldspar but <75% Quartz. Coarse grains, indicates tectonically active region. Rapid burial close to the source.
3. Graywacke: Very immature, lots of clays mixed in, poorly sorted, angular, indicates little transport and rapid burial.
4. Lithic: <75% Quartz, not as much feldspar, other minerals mixed in, common in deltas and marshes.

Weathering Classifications:

Mature: well rounded, well sorted, and long transport. Unit has undergone rigorous weathering. Ferromags and Feldspars are gone.
Immature: not rounded or less rounded. Poorly sorted, the unit still has some ferromags and feldspars left. Has not undergone rigorous weathering.

Rocks as a Collective Unit:

Earth Scientists also look at how the rock units collectively fit together. The rock units are often distinguished by color such as Red Rock. Red Rock comes from the source and ends up in the environment (i.e. marsh, lake, etc.). The red color is generally caused by oxidation of ferric (iron) materials in the rock.

Lithification: the process of becoming rock, (i.e. the Red Rock in Monument Valley, California.) The depth of the rock represents time. The rock will continue to be laid down as long as there is no change in the environment of deposition.

Rock Unit: distinctive for it's lithology and structure. Indicates all were laid down in the same environment and at the same time.

Formation: rock that is lithologically distinctive (same sediment type throughout). A formation has recognizable contacts above and below and can be traced over distances. Formations are mapable units.

Lithologically Distinctive: clearly the same rock with grain size, color, and composition all the same throughout. However, it may or may not be the same age everywhere.

Rock Formations have two given names that reflect their geographic locale and lithological type. For example: Bright Angel Shale. This formation is found in the Grand Canyon.

Skill 18.6 Identify and describe characteristics of various rock types and minerals and how they are formed.

Minerals are natural inorganic compounds. They are solid with homogenous crystal structures. Crystal structures are the 3-D geometric arrangements of atoms within minerals. Though these mineral grains are often too small to see, they can be visualized by X-ray diffraction. Both chemical composition and crystal structure determine mineral type. The chemical composition of minerals can vary from purely elemental to simple salts to complex compounds. However, it is possible for two or more minerals to have identical chemical composition, but varied crystal structure. Such minerals are known as polymorphs. One example of polymorphs demonstrates how crystal structures influence the physical properties of minerals with the same chemical composition: diamonds and graphite. Both are made from carbon, but diamonds are extremely hard because the carbon atoms are arranged in a strong 3-D network while graphite is soft because the carbon atoms are present in sheets that slide past one another. There are over 4,400 minerals on Earth, which are organized into the following classes:

Silicate minerals are composed mostly of silicon and oxygen. This is the most abundant class of minerals on Earth and includes quartz, garnets, micas, and feldspars.

Carbonate class are formed from compounds including carbonate ions (including calcium carbonate, magnesium carbonate, and iron carbonate). They are common in marine environments, in caves (stalactite and stalagmites), and anywhere minerals can form via dissolution and precipitation. Nitrate and borate minerals are also in this class.

Sulfate minerals contain sulfate ions and are formed near bodies of water where slow evaporation allows precipitation of sulfates and halides. Sulfates include celestite, barite, and gypsum.

Halide minerals include all minerals formed from natural salts including calcium fluoride, sodium chloride, and ammonium chloride. Like the sulfides, these minerals are typically formed in evaporative settings. Minerals in this class include fluorite, halite, and sylvite.

Oxide class minerals contain oxide compounds including iron oxide, magnetite oxide, and chromium oxide. They are formed by various processes including precipitation and oxidation of other minerals. These minerals form many ores and are important in mining. Hematite, chromite, rutile, and magnetite are all examples of oxide minerals.

Sulfide minerals are formed from sulfide compounds such as iron sulfide, nickel iron sulfide and lead sulfide. Several important metal ores are members of this class. Minerals in this class include pyrite (fool's gold) and galena.

Phosphate class includes not only those containing phosphate ions, but any mineral with a tetrahedral molecular geometry in which an element is surrounded by four oxygen atoms. This can include elements such as phosphorous, arsenic, and antimony. Minerals in this class are important biologically, as they are common in teeth and bones. Phosphate, arsenate, vanadate, and antimonite minerals are all in this class.

Element class minerals are formed from pure elements, whether they are metallic, semi-metallic or non-metallic. Accordingly, minerals in this class include gold, silver, copper, bismuth, and graphite as well as natural alloys such as electrum and carbides.

Rocks

Rocks are simply aggregates of minerals. Rocks are classified by their differences in chemical composition and mode of formation. Generally, three classes are recognized: igneous, sedimentary, and metamorphic. However, it is common that one type of rock is transformed into another and this is known as the rock cycle.

Igneous rocks are formed from molten magma. There are two types of igneous rock: volcanic and plutonic. As the name suggests, volcanic rock is formed when magma reaches the Earth's surface as lava. Plutonic rock is also derived from magma, but it is formed when magma cools and crystallizes beneath surface of the Earth. Thus, both types of igneous rock are magma that has cooled either above (volcanic) or below (plutonic) the Earth's crust. Examples of this type of rock include granite and obsidian glass.

Sedimentary rocks are formed by the layered deposition of inorganic and/or organic matter. Layers, or strata, of rock are laid down horizontally to form sedimentary rocks. Sedimentary rocks that form as mineral solutions (i.e., seawater) evaporate and are called precipitate. Those that contain the remains of living organisms are termed biogenic. Finally, those that form from the freed fragments of other rocks are called clastic. Because the layers of sedimentary rocks reveal chronology and often contain fossils, these types of rock have been key in helping scientists understand the history of the earth. Chalk, limestone, sandstone, and shale are all examples of sedimentary rock.

Metamorphic rocks are created when rocks are subjected to high temperatures and pressures. The original rock, or protolith, may have been igneous, sedimentary or even an older metamorphic rock. The temperatures and pressures necessary to achieve transformation are higher than those observed on the Earth's surface and are high enough to alter the minerals in the protolith. Because these rocks are formed within the Earth's crust, studying metamorphic rocks gives us clues to conditions in the Earth's mantle. In some metamorphic rocks, different colored bands are apparent. These are the result of strong pressures applied from specific directions and are termed foliation. Examples of metamorphic rock include slate and marble.

COMPETENCY 19.0 UNDERSTAND AND APPLY KNOWLEDGE OF THE HISTORICAL EVOLUTION OF EARTH'S FEATURES AND LIFE FORMS THROUGH GEOLOGIC TIME.

Skill 19.1 Recognize the scope of geologic time by distinguishing between the human time scale and the geologic time scale.

James Hutton, a Scottish gentleman farmer, medical doctor, and amateur geologist, first postulated in 1785, the primary principle upon which modem geology is based; the **Principle of Uniformitarianism**. Hutton's observations led him to conceptualize that there is a uniformity of geologic processes, past and present. In essence, the Principle of Uniformitarianism states that the geologic processes that are happening today also happened in the past, and that the present is the key to the past.

Hutton's simple but profound concept provided science with a dynamic tool with which to view the past. Suddenly, when viewed in light of Hutton's principle, logical explanations were available for a wide variety of questions that had stymied geologists. By implication, the corollary to the principle of Uniformitarianism is that, because the processes that modified our planet in the past continue to work at the slow pace of today, an immense period of time must pass in order for the processes to accomplish the task. Thus by extension, Hutton's observations also introduced the concept of Geologic Time. For example, the period of time required to raise a mountain thousands of feet above the surface and then wear it down again had to have taken place over a truly impressive period of time, on the order of hundreds of thousands of human lifetimes.

Skill 19.2 Identify methods and technologies used to study Earth's history (e.g., relative and absolute dating techniques) leading to the divisions of geologic time.

Time
There are two basic concepts involved in measuring time: relative and absolute time. Although both concepts are used in the Earth Sciences, absolute time provides the most accurate concept of describing the passage of time.

Relative time- making a comparison of the order of one object relative to the other. Example: This object is older than that object.

Absolute time- the exact age of an object.
Example: This object is xxx years old.

In Earth Science, time is measured using Relative and Absolute Dating techniques.

Relative Dating

As the name implies, this technique involves a comparison of rock assemblages, through application of the principle laws of geology.

Absolute Dating

Definition: Assigning time in years before present to a rock or fossil assemblage. Four methods are used in absolute dating: Dendrochronology, Varves, Radiometric Dating, and Fission Tracking Dating.

Dendrochronology: tree ring dating. Accuracy range restricted to past 6,000 years. This was an early technique used in absolute dating. By measuring the width of the tree rings and counting the number of rings, a scientist could make an approximation of the climatology of the area and how old the tree was when it died.

Varves: sea bands in a lake. Maximum dating range limited to the past 20,000 years. Both Dendrochronology and Varves are relatively inaccurate, and are poor methods of absolute dating, and neither method is used much anymore.

Radiometric Dating: The **most accurate** method of absolute dating, this technique measures the decay of naturally occurring radioactive isotopes. These isotopes are great timekeepers because their rate of decay is constant. Elements decay because of the inherent structure of the nucleus of the atoms. Neutrons hold the positively charged protons together. However, the positive protons attempt to repel each other. In some heavy elements, the protons repel each other to such a degree that the proton tears itself apart (decays) and by losing protons, becomes another element. The decay starts the moment an isotope crystallizes in a rock unit, and chemicals, weathering, environment, or temperature does not affect the rate of decay.

The radioactive decay causes the (mother) element to change into an (daughter) element. The Mother-Daughter relationship of produced nuclides during the series of isotope decay is the basis for radiometric dating. Although many isotopes are used in radiometric dating, the most widely known method is referred to as **Carbon-14 dating**. Carbon-14 is unstable and decays, decomposes and transmutes to Carbon-12. The dating process compares the ratio of Carbon-14 to Carbon-12 in an object. Since the decay occurs at a known rate, it is very predictable and can be used as a clock standard. However, Carbon-14 decays quickly and can only be used to date organic compounds less than 40,000 years old.

Knowing the half-life of the isotopes is the key factor in the radiometric dating process. If we know the half-life, we can compare the ratio of isotopes found in the object, and count backward to get an accurate date. The most common element checked is the ratio of Uranium to Lead. Example: 1 gram of 238Uranium. After 100 million years, you have 0.013g of 206Pb (lead) and 0.989 of 238U. After 4.5 billion years, you have 0.433g 206Pb and .500g of 238U. Therefore, the half-life of 238U is roughly 4.5 billion years.

Note: Only Carbon-14 can be used to date organic compounds. The other isotopes are not found in organic compounds.

Radiometric Isotopes Commonly Used in Absolute Dating

Mother Isotope	Daughter Isotope	Half-life (in years)	Dating Range (in years)
40K (Potassium)	10Ar (Aragon)	1.3 billion	>10 million
87 Rb (Rubidium)	87St (Strontium)	47 billion	>100 million
235U (Uranium)	407U (Uranium)	700 million	>1 million
14C (Carbon)	206 Pb (Lead)	5,730	>750<40,000

Fission Tracking Dating

Fission tracking dating is a new alternative method that doesn't rely on mother-daughter techniques. Instead, it is based on counting the scars caused by the collision of atoms. There is a gap in the radiometric-dating scheme between 1 mya to 50 ya. Fission tracking fills in this gap. However, because the method is relatively new, it is controversial and not completely accepted by all scientists.

Dating Non-organics

Radioactive isotope dating is a relatively new invention. Prior to its invention, geologists relied upon relative dating methods. Radioactive isotopes are good for dating rock units because they are common in igneous rocks, some metamorphic rocks, and occasionally in sedimentary rocks such as sandstone. However, the presence of the isotope in sandstone can tell us how old the material is, but not when it was laid down. To determine the age of the sandstone, scientists try to absolute date layers above and below the sandstone.

Thus, dating sandstone uses a combination of absolute and relative dating methods. You can only tell if the lay down occurred between two dates, not the absolute date of the event.

Fossils found in the sandstone are often not directly dated. Instead, the rock units surrounding it are dated. Earth Scientists examine the layers above and below the fossil to get an approximate range of how old the fossil must have been when it was laid down in the rock layers.

Chronostratagraphic Unit: rock units that correspond to a geologic or geochronologic event.

4. Geologic Time Scale

In the Earth Sciences, when you talk about time, you must think in terms of huge expanses of time.

Geological Time Scale: the calendar/clock of events in geology based on the appearance and disappearance of fossil assemblages.

The scale is divided into time units that are given distinctive names and approximate start and stop dates. These dates are based upon a reexamination of previously discovered fossils, using absolute dating techniques.

Eons: The largest scale division
Eras: Divided into sub categories based on profound differences in fossil life.
Periods: Smaller divisions within eras based on less profound differences in the fossil record.
Epochs: Sub-categories within some periods that are specific to the types of fossils found within.
Pre-Cambrian Time: Comprised of the Hadean, Archean, and Proterozoic Eons, 87% of all geologic time is considered Pre-Cambrian.

Major Geological Time Scale Divisions

Eon, Era, Period, or Epoch	Time Start (mya)
Hadean Eon	4600
Archean Eon	3800
Early Era	3800
Middle Era	3400
Late Era	3000
Proerozoic Eon	2500
Paleo-proterozoic Era	2500
Meso-proterozoic Era	1600
Neo-proterozoic Era	1000
Phanerozoic Eon	570
Paleozoic Era	570
Cambrian Period	570
Ordovician Period	505
Silurian Period	438
Devonian Period	408
Carboniferous Period, includes the:	360
Mississippian Period	360
Pennsylvanian Period	320
Permian Period	286
Mesozoic Era	245
Triassic Period	245
Jurassic Period	208
Cretaceous Period	144
Cenozoic Era	66
Tertiary Period, comprised of:	66
Paleogene Period	66
Paleocene Epoch	66
Eocene Epoch	58
Oligocene Epoch	37
Neogene Period	24
Miocene Epoch	24
Pliocene Epoch	5
Quartenary Period	2
Pleistocene Epoch	2
Holocene Epoch	10,000 years

Skill 19.3 Analyze evidence for hypotheses and theories regarding Earth's origins and the evolution of Earth's features and life forms (e.g. paleontology, paleoclimatology, paleogeology).

It is estimated that the earth is 4.54 billion years old. Rocks on every continent have been found to be older than 3.5 billion years through radiometric dating. The oldest rocks are the Acasta Gneisses in NW Canada that measure to be 4.03 billion years old. There are also individual crystals found in Australia that have been measured to be 4.3 billion years old. However, the earth is thought to be older than these oldest rocks, because these rocks were made from lava flows and/or sediment. This indicates that these rocks must have been formed from other rocks destroyed (melted) by Earth processes.

There is significant evidence that indicates the evolution of earth's features and life forms. This evidence is found in paleontology, paleoclimatology, and paleogeology.

Evidence of earth's features and life forms in paleontology

The changes in Earth's life forms can be seen implicitly through the study of paleontology. By studying fossils found in sedimentary layers of the earth, scientists can put together a comprehensive list of the species that existed during various times in earth's history. As scientists have found, the diversity of species has increased as time has progressed. Although mass extinctions have been discovered in the rock record, evidence shows that organisms have become more diverse and more advanced over time.

The study of paleontology can also help scientists piece together the evolution of Earth's features. Alfred Wegener, the father of the theory of plate tectonics, looked at the distribution of fossils over the planet and found patterns in the rock record. He was particularly interested in Mesosaurus fossils. The Mesosaurus was a freshwater reptile whose remains can be found in South America and in Africa. Wegener believed that the Mesosaurus did not swim across the Atlantic in order to be found on both continents (as it was a freshwater reptile), but that the Mesosaurus was able to walk or swim when the two continents were connected. The Glossopteris, a tropical fern, was also a fossil that Wegner studied. The distribution of the Glossopteris fossils was widespread and includes several continents that are no longer suitable for tropical plants. This fact led Wegener to the conclusion that the continents must have been moved to their current position and were once connected.

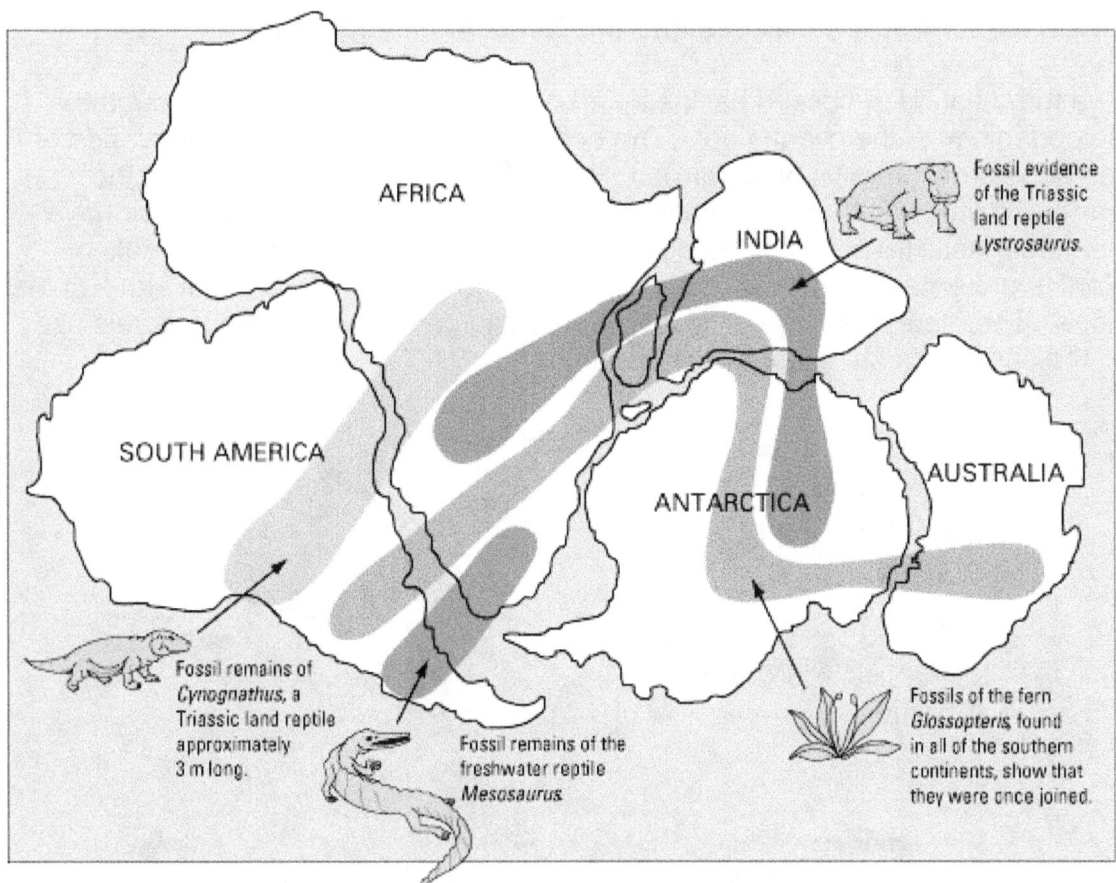

Figure: The distribution of indicator fossils across continents. This evidence was used in the continental drift theory.

Evidence for earth's features and life forms in paleoclimatology

Paleoclimatology is the study of Earth's ancient climates. Through studying layers of ice and rock, the climate of the earth during Earth's history can be determined. Pollen trapped in layers of ice (whose age has been determined through interferometry) can give information on the total amount of plant life on the earth at a given time. Scientists have also drilled ice cores in order to analyze air bubbles trapped in the ice. These air bubbles are samples of the ancient atmosphere when they were trapped. Through analysis of the ratio of oxygen isotopes, scientists can determine the average temperature of the earth at that time, and therefore determine what organisms might have thrived. By studying sedimentary layers and the fossils therein, large shifts in climate can be determined.

The formation of mountains and plate tectonics can also be proven using paleoclimatology. Tropical fossils have been found in rock on snow topped mountains. This explains that earth's mountains and other features have been formed through plate tectonic processes and the fossils were pushed up through collision of continents.

Evidence for Earth's features and life forms in paleogeology

The formation of the ocean basins as an earth feature is explained through the paleogeology of the ocean floor. The ocean floor has a ridge. On either side of the ridge, iron minerals have formed, lining up with the magnetic pole of the earth. By studying these minerals, paleogeologists have found that there have been magnetic reversals of the poles. The evidence of magnetic reversals of Earth's poles has been seen at equal distance from the ridge on both sides of the ridge. Therefore, the ridge must be a spreading center where iron minerals rise to the surface, pushing the plates apart.

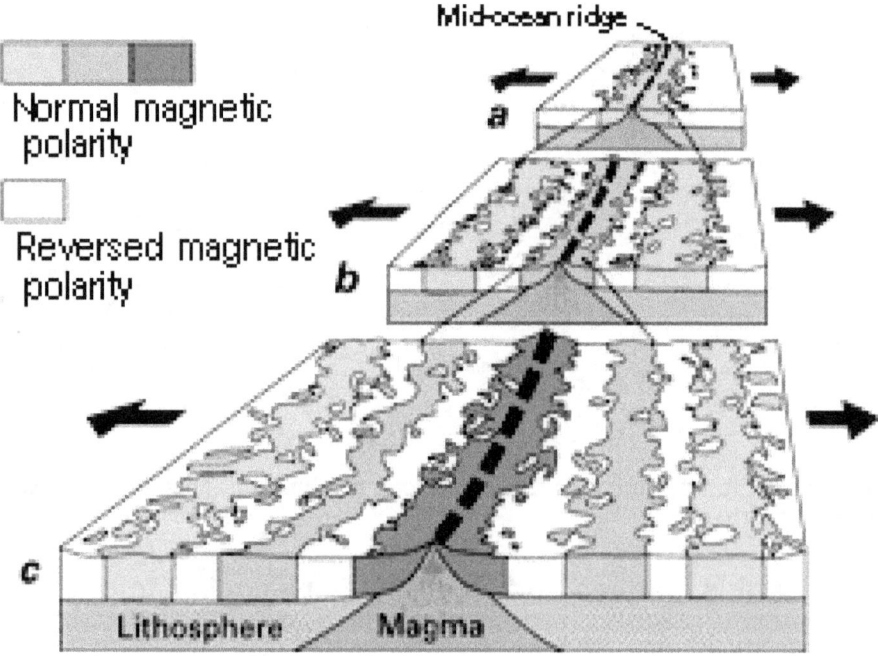

Figure: Magnetic reversals at the mid-ocean ridge.

Paleogeologists have also discovered that mountain ranges in eastern North America line up with mountain ranges in the UK and Scandinavia, signifying that these areas were once connected.

It is through the study of paleontology, paleoclimatology, and paleogeology that scientists have discovered the evolution of life and earth features.

Skill 19.4 Describe how rock strata and fossils lead to inferences about geological environments and climatic conditions.

The Earth's materials-rocks, soils, and sediments-are piled upon each other in layers called **strata**. Understanding the relative orientation and arrangement of the strata provides important information about the Earth's history and the ongoing sequence of events and processes that helped shape that history.

Certain assumptions come into play when determining the sequence of events and correlating stratigraphic layers. These assumptions are the basis for the Principles of Geology. Note: The basic geologic principles are often referred to as "Laws."

THE BASIC PRINCIPLES (LAWS) OF GEOLOGY

Principle of Uniformitarianism: Processes that are happening today also happened in the past.

Principle of Cross-Cutting Relations: A rock is younger than any rock it cuts across.

Principle of original Horizontality: Rock units are originally laid down flat. Something happened to cause them to change orientation.

Principle of Inclusion: Any rock enclosed within another rock must be older than the rock in which it is enclosed.

Principle of Original Lateral Continuity: A rock unit extends out in all directions.

Principle of Super Position: The rock on the bottom is older than the rock on top.

Principle of Biologic Succession: Fossils correspond to particular periods of time.

Stratigraphy: the study of regional landforms with view to detailing and understanding the sequence of events and relative timeframe in which those events occurred within the regions.

Sequence of Events: To understand the basic sequence of events for a particular landform, you must think in relative terms. In other words, something happened before or after to cause the event. At this point, don't be overly concerned about absolute dates, but concentrate instead on determining a sequence of events of a landform's strata in terms relative to each other.

Environment of Deposition:

Earth's History is contained in its rocks. And although all types of rocks are important in constructing the historical record, sedimentary rocks are particularly useful because their current structure and composition provides important information about their component's original structure and composition. However, simply recognizing the structure and composition is not enough to successfully reconstruct the historical record. Equally important is knowing the Environment of Deposition (EOD): the physical conditions present at the time the rock was formed.

Because a particular environment has distinctive features, by examining sedimentary rocks we can find clues to better understand the historical record of the Earth's original structure and composition. To determine the environment of deposition, scientists initially look at the Color, Texture, and Sedimentary Structures of Rocks.

Color:

> Black, Dark Gray, or Dark Green: a reducing environment. No oxygen was present when laid down. A reducing environment is typically rich in collected organic matter because the lack of oxygen impedes the decay of organic matter.
>
> EOD = Swamp, tidal pool, lagoon, restricted basins, or the bottom of a lake.
>
> Red, Orange, Yellow, Brown, Bright Green: An oxidizing environment. Lots of oxygen was present when laid down.
>
> EOD = Beach, stream, flood plain deposits, and sometimes, deltas or parts of deltas.
>
> Texture: The size, shape, and arrangement of the mineral grains within the sedimentary rock.

Besides looking at the individual granular properties, Earth Scientists also look at rocks as a whole unit, how the rocks fit together as a collective unit, and in what sequence the sedimentary units were laid down.

COMPETENCY 20.0 UNDERSTAND AND APPLY KNOWLEDGE OF THE TRANSFER OF ENERGY AND CYCLING OF ELEMENTS AND COMPOUNDS IN EARTH SYSTEMS.

Skill 20.1 Identify chemical reservoirs for various elements and compounds (e.g., carbon, nitrogen, water).

A chemical reservoir is a place where elements and chemical compounds are held for a long period of time.

Carbon reservoir

The four major reservoirs for carbon are the atmosphere, the terrestrial biosphere, the oceans, and sediments. In the atmosphere, carbon exists primarily as carbon dioxide. In the terrestrial biosphere, carbon is found in living organisms, freshwater systems, and in the soil as decaying organisms. The ocean reservoir for carbon includes dissolved carbon from weathering of rocks, marine organisms, and tests or shells of marine organisms ($CaCO_3$). The sediment reservoir for carbon includes fossil fuels and limestone.

Nitrogen reservoir

The chemical reservoirs for nitrogen include the atmosphere, ocean, terrestrial biomass, and soil. The atmosphere is the largest reservoir for nitrogen. Nitrogen exists as N_2, N_2O, and fixed nitrogen in the atmosphere. Lightning fixes nitrogen in the atmosphere. In the oceans, nitrogen is found in dissolved minerals, in living organisms, and in sediments. The soil reservoir for nitrogen includes organic material, ammonium, and decaying matter. Nitrogen also exists to a large extent in living tissue.

Water reservoir

Water exists everywhere. The largest chemical reservoir for water is the oceans. Although the majority of the earth is covered in water, only a small proportion is accessible for human use.

Oxygen reservoir

Oxygen is found in the lithosphere, biosphere, and atmosphere. Oxygen in the lithosphere exists in minerals, primarily silicates such as quartz. Oxygen exists as a chemical compound in the cells of living organisms in the biosphere. The atmosphere reservoir for oxygen occurs mainly through the photosynthesis exchange of oxygen. Photolysis, the breakdown of water and nitrites into their component parts via UV radiation, also releases oxygen into the atmosphere.

Skill 20.2   Trace the cycle of various elements and compounds through the lithosphere, hydrosphere, biosphere, and atmosphere.

The Nitrogen Cycle

The nitrogen cycle has 5 major steps.

Step 1: Nitrogen fixation. Nitrogen fixation, the conversion of nitrogen into a form that living things can use, involves changing N_2 in the atmosphere (gaseous nitrogen) into ammonia (NH_3) or nitrate (NO_3^-) in the lithosphere. Gaseous nitrogen becomes fixed as nitrate through volcanic action, lightning, and combustion. The majority of the nitrogen becomes fixed as ammonia through biological means, however. Nitrogen- fixing bacteria and cyanobacteria living in the soil and hydrosphere aquatic environments use an enzyme called nitrogenase to break down gaseous nitrogen and bond it with hydrogen.

Step 2: Nitrification. In nitrification, soil bacteria converts ammonia to nitrate.

Step 3: Assimilation. Assimilation is the process by which plant roots absorb nitrate and ammonia and integrate them in their nucleic acids and plant proteins. This assimilation transports the nitrogen to the biosphere. As animals consume the plants, the animals assimilate the nitrogen.

Step 4: Ammonification. The process of ammonification is when ammonifying bacteria converts the nitrogen found in animal waste products and decomposing organisms into ammonia. This ammonia can undergo nitrification and assimilation once again.

Step 5: Denitrification. Denitrifiying bacteria takes nitrate and returns it to the atmosphere.

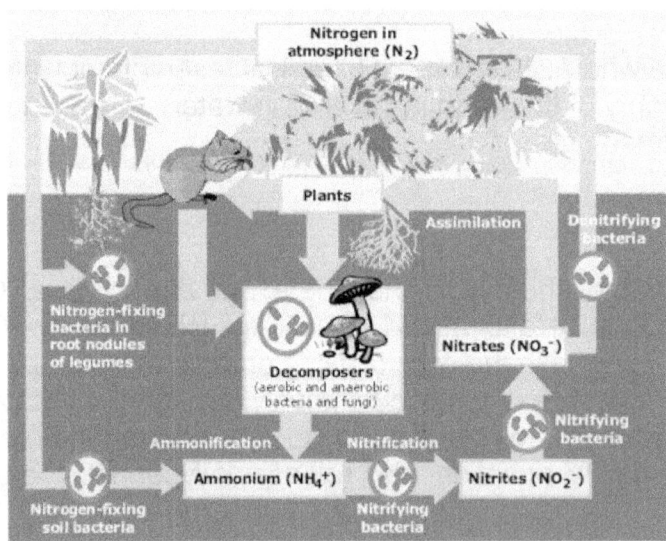

The Water Cycle

Water in the hydrosphere (rivers, oceans, lakes, etc) evaporates into the atmosphere or sublimates from icecaps. The water vapor condenses to form clouds. Precipitation transports the water back to the lithosphere and the hydrosphere. Some of the water is stored in snow and ice, until it melts and becomes runoff or percolates into the ground. Water also infiltrates into the bedrock to become groundwater or water used by plant life. The process of evapo-transpiration in plants returns the water back to the atmosphere. The groundwater is stored for long periods of time. Groundwater eventually discharges into streams and oceans, becoming available for evaporation. Precipitation can also run off into streams and lakes, to start the cycle once more.

The Carbon Cycle

Plants, algae and cyanobacteria in the biosphere remove carbon dioxide from the air and integrate it into sugars. The sugars are used as a source of energy for the manufacture of the sugar, the animal that consumes them, or by a decomposer. Cellular respiration returns carbon dioxide to the atmosphere. In aquatic environments or the hydrosphere, algae, cyanobacteria, and plants take dissolved carbon dioxide in the water, and perform cellular respiration, returning the carbon dioxide to the atmosphere. Another way that carbon dioxide is returned to the atmosphere is through the combustion of fossil fuels stored in the lithosphere. By oxidizing the carbon in the fossil fuels, carbon dioxide is released to the atmosphere. The carbon dioxide that exists in the atmosphere cycles to be used in photosynthesis once more.

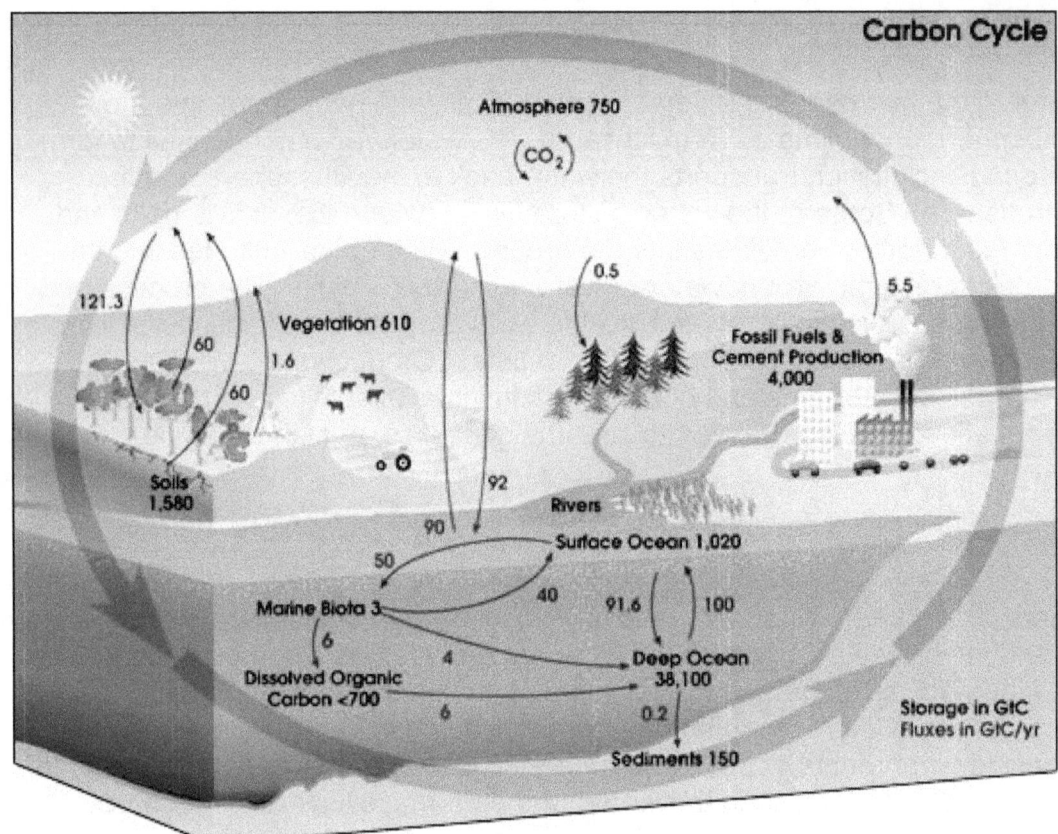

The carbon cycle. The number in black is the billion tons of carbon stored in the system. The number in purple is the billion tons of carbon in flux each year.

Skill 20.3 **Analyze the factors and interpret data that indicate the transfer of energy within and among Earth's land, water, and atmospheric systems (e.g., ocean currents and temperatures, surface albedo, atmospheric circulation).**

The majority transfer of energy begins with the sun. The sun provides 17.3×10^{16} watts of energy to earth's systems. This is 5000 times more energy than the earth receives from its interior. The surface albedo, or the amount of energy that gets reflected back into space from ice and clouds, is 5.28×10^{16} watts. The remaining energy from the sun, and the energy from the interior of the earth, drives earth's systems.

Ocean Currents and Temperatures

Most of the earth's energy at the surface is in the earth's oceans. This is primarily because water's heat capacity is 5 times that of land. Some of this heat gets radiated back into the atmosphere. Some of this heat drives the ocean currents. The ocean water at the equator is a higher temperature than the ocean water temperature at the poles due to the angle at which the solar energy hits the ocean. Due to this uneven distribution of heat in the oceans, convection currents begin to form. As wind begins to move over the surface of the water, drag pulls along the heated water (as heated water stays at the surface). The Coriolis Effect rotates the currents in their respective hemispheres.

An example of a global ocean current is the Gulf Stream. Water in the Atlantic Ocean near the equator gets heated by the sun's energy. The Coriolis effect drives the warmed water north. When it reaches the North Atlantic is cools and sinks. It flows along the bottom of the Atlantic Ocean to the South Atlantic and into the Indian Ocean. In the Indian Ocean and Pacific Ocean, the water absorbs energy from overlying air masses, heats, rises, and then feeds back into the Gulf Stream.

Atmospheric Circulation

Uneven heating of the earth's surface causes the earth's atmosphere to be warmed unevenly. Air over the equator rises. As it rises, it moves northward, cools over the poles and sinks. When it sinks, it moves southward, warms and begins the convection current again. The Coriolis effect causes the wind to veer in each hemisphere, causing three circulation belts in each hemisphere to form.

The Hydrologic Cycle

The hydrologic cycle moves energy between the hydrosphere and the atmosphere. The water at the earth's surface absorbs energy from the sun. The water evaporates and the warmed, humid air rises. As it rises, it cools. The energy is then released as the water condenses to form cloud. The water eventually falls as precipitation and once again absorbs heat from the sun, starting the energy transfer system once again.

Skill 20.4 Describe the interrelationships among land, water, and atmospheric systems and how these relationships explain various natural processes, cycles, and events (e.g., weather and climate systems, El Niño events).

El Niño refers to a sequence of changes in the ocean and atmospheric circulation across the Pacific Ocean. The water around the equator is unusually hot every two to seven years. Trade winds normally blow east to west across the equatorial latitudes, piling warm water into the western Pacific. A huge mass of heavy thunderstorms usually forms in the area and produces vast currents of rising air that displace heat poleward. This helps create the strong mid-latitude jet streams. The world's climate patterns are disrupted by this change in location of thunderstorm activity.

Air masses moving toward or away from the Earth's surface are called air currents. Air moving parallel to Earth's surface is called **wind**. Weather conditions are generated by winds and air currents carrying large amounts of heat and moisture from one part of the atmosphere to another. Wind speeds are measured by instruments called anemometers.

The wind belts in each hemisphere consist of convection cells that encircle Earth like belts. There are three major wind belts on Earth: (1) trade winds (2) prevailing westerlies, and (3) polar easterlies. Wind belt formation depends on the differences in air pressures that develop in the doldrums, the horse latitudes, and the polar regions. The Doldrums surround the equator. Within this belt heated air usually rises straight up into Earth's atmosphere. The Horse latitudes are regions of high barometric pressure with calm and light winds and the Polar regions contain cold dense air that sinks to the Earth's surface.

Winds caused by local temperature changes include sea breezes, and land breezes.

Sea breezes are caused by the unequal heating of the land and an adjacent, large body of water. Land heats up faster than water. The movement of cool ocean air toward the land is called a sea breeze. Sea breezes usually begin blowing about mid-morning; ending about sunset.

A breeze that blows from the land to the ocean or a large lake is called a **land breeze.**

Monsoons are huge wind systems that cover large geographic areas and that reverse direction seasonally. The monsoons of India and Asia are examples of these seasonal winds. They alternate wet and dry seasons. As denser cooler air over the ocean moves inland, a steady seasonal wind called a summer or wet monsoon is produced.

The air temperature at which water vapor begins to condense is called the **dew point.**

Relative humidity is the actual amount of water vapor in a certain volume of air compared to the maximum amount of water vapor this air could hold at a given temperature.

Water that falls to Earth in the form of rain and snow is called **precipitation.** Precipitation is part of a continuous process in which water at the Earth's surface evaporates, condenses into clouds, and returns to Earth. This process is termed the **water cycle**. The water located below the surface is called groundwater.

The impacts of altitude upon climatic conditions are primarily related to temperature and precipitation. As altitude increases, climatic conditions become increasingly drier and colder. Solar radiation becomes more severe as altitude increases while the effects of convection forces are minimized. Climatic changes as a function of latitude follow a similar pattern (as a reference, latitude moves either north or south from the equator). The climate becomes colder and drier as the distance from the equator increases. Proximity to land or water masses produce climatic conditions based upon the available moisture. Dry and arid climates prevail where moisture is scarce; lush tropical climates can prevail where moisture is abundant. Climate, as described above, depends upon the specific combination of conditions making up an area's environment. Man impacts all environments by producing pollutants in earth, air, and water. It follows then, that man is a major player in world climatic conditions.

Skill 20.5　Identify factors, including human activities, that affect the cycling of elements and compounds in Earth systems (e.g., volcanic eruptions, agriculture practices, logging, fossil fuels).

An important topic in science is the effect of natural disasters and events on society and the effect human activity has on inducing such events. Naturally occurring geological, weather, and environmental events can greatly affect the lives of humans. In addition, the activities of humans can induce such events that would not normally occur.

Nature-induced hazards include floods, landslides, avalanches, volcanic eruptions, wildfires, earthquakes, hurricanes, tornadoes, droughts, and disease. Such events often occur naturally, because of changing weather patterns or geological conditions. Property damage, resource destruction, and the loss of human life are the possible outcomes of natural hazards. Thus, natural hazards are often extremely costly on both an economic and personal level.

While many nature-induced hazards occur naturally, human activity can often stimulate such events. For example, destructive land use practices such as mining can induce landslides or avalanches if not properly planned and monitored. In addition, human activities can cause other hazards including global warming and waste contamination. Global warming is an increase in the Earth's average temperature resulting, at least in part, from the burning of fuels by humans. Global warming is hazardous because it disrupts the Earth's environmental balance and can negatively affect weather patterns. Ecological and weather pattern changes can promote the natural disasters listed above. Finally, improper hazardous waste disposal by humans can contaminate the environment. One important effect of hazardous waste contamination is the stimulation of disease in human populations. Thus, hazardous waste contamination negatively affects both the environment and the people that live in it.

TEACHER CERTIFICATION STUDY GUIDE

COMPETENCY 21.0 UNDERSTAND AND APPLY KNOWLEDGE OF THE STRUCTURE AND PROCESSES OF THE HYDROSPHERE AND ATMOSPHERE.

Skill 21.1 Describe the physical and chemical nature of water and its influence on the lithosphere, hydrosphere, biosphere, and atmosphere.

While the hydrosphere, lithosphere, and atmosphere can be described and considered separately, they are actually constantly interacting with one another. Energy and matter flows freely between these different spheres. For instance, in the water cycle, water beneath the Earth's surface and in rocks (in the lithosphere) is exchanged with vapor in the atmosphere and liquid water in lakes and the ocean (the hydrosphere).

Water (H_2O) is significantly different from its immediate Hydrogen compound cousins. **Compounds** are substances that contain two or more elements in a fixed proportion. Generally, the heavier molecules have higher boiling and freezing states based upon molecular weight.

Example: H_2Te = Hydrogen Tellide.

A group of atoms held together by chemical bonds is called a **molecule**. The bonds form when the small, negatively charged electrons found near the outside of an atom are shared or transferred between the atoms. The bonds formed by the shared pair of electrons are known as **covalent bonds**.

Most substances tend to adopt a solid or gaseous form. Water is different. It wants to be a liquid. A water molecule forms when covalent bonds are established between two hydrogen atoms and one oxygen atom.

However, unlike the other hydrogen compounds, water has its two hydrogen molecules on one side of the atom. It's a polar molecule.

This arrangement of atoms is based on the distribution of the water molecule's oxygen electrons. The electrons cause the geometric shape of the molecule to be angular. This angular shape makes the molecule electrically asymmetrical (polar).

This polar arrangement gives water some very special properties:

- The polar molecule acts similar to a magnet. Its positive ends (the hydrogen) attracts particles having a negative charge, and its negative end (the oxygen) attracts particles that have a positive charge. This arrangement is the basis for the water molecule's rightful description as the **"Universal Solvent."** When water comes into contact with compounds- for example salts- held together by the attraction of opposite charges, the water molecule separates that compound's elements from each other.

- Another unique property of water is that water likes itself; it has a natural tendency to stick to itself. Once again, this property is based upon the polar nature of the water molecule. It attracts other water molecules. When the molecules stick together, they are attached through Hydrogen Bonds, giving the molecule a property called **cohesion**. Cohesion gives water an unusually strong surface tension, and its capillary action makes the water spread. When the water spreads, **adhesion**, the tendency of water to stick to other materials, allows water to adhere to solids, making them wet.

Water is the only known substance that readily exists in all three states of matter: liquid, solid, and gas. The hydrogen bonds holding water together are important. If they didn't exist, water would fly apart to form a gas. However, water primarily wants to be a liquid, and its liquid state range is meaned at 16°C.

Ocean water is composed of additional elements, mainly salts.

Skill 21.2   Demonstrate knowledge of the characteristics of oceans (e.g., currents, tides, nutrient availability) and freshwater systems (e.g., lakes, aquifers, glaciers).

98% of the water on Earth is found in the oceans. The remaining 2% is either locked up in glaciers as land ice, found in the ground (groundwater), found in rivers and lakes, or is present in the atmosphere.

Around 0.6 billion years ago the ocean stabilized into its present chemical composition. In every kilogram of seawater there is 965g freshwater and 35g inorganic salts.

Surface area of the Earth covered by oceans totals 361,100,000 km. The volume of the surface is 1,370,000,000 km. Its average depth is 3,796 meters. The ocean has multiple layers that can be compared to the layers of an onion. Like an onion, the various layers overlap each and meet at distinctive boundaries.

Central Waters: Includes the shallow surface waters. These basically go around and around.

EARTH & SPACE SCIENCE

Intermediate Waters: Can go both directions. Rarely found past a depth of 900 meters.

Deep Water.

Bottom Water.

All the different layers of circulation are eventually tied together at the CDW (Circumpolar Deep Water), which has its origin in the North Atlantic.

Average temperature of the oceans is 3.9 °C (39.0 °F). The average salinity of ocean water is 34.482 grams per kg. The most abundant elements in ocean water are:

> Oxygen = 86%
> Hydrogen = 11 %
> Chlorine = 1.9%
> Sodium = 1.1 %
> Magnesium = 0.1%

Glaciers

Glacier: a large, long-lasting mass of ice formed on the land that moves due to gravity. A glacier forms as the result of snow accumulating, compacting, and crystallizing over a period of years.

Glaciated Terrain refers to areas that have been or are still currently affected by glaciers. There are two types of terrain: **Alpine**-found in mountainous regions, and **Continental**- areas where a large part of the continent is covered by glacial ice.

Glaciers can be found in both cold, Polar, regions and more temperate areas that meet the proper conditions for glaciations.

The area must either be cold enough so that the accumulation does not melt away during the summer months, such as in the Polar Regions, or, the area must receive enough snowfall during the winter months so that during the summer, the volume of snow is simply too much to completely melt away.
Example: The mountainous regions of Washington State, or the High Andes in South America.

The majority of glaciations occur in the colder areas such as the Polar Regions. Because little melting occurs at any time during the year, significant accumulations of snow and ice remain present.

Glaciers cover about 10% of the Earth's surface, and 85% of that total area is comprised of the glaciations present in Antarctica.

Rivers

Streams: a body of water confined to a channel. **Regardless of size**, geologists call all moving bodies of water confined to a channel "streams".

Hydrologic Cycle: the Earth's water balance. All the water on our planet is all that there is available. It changes location and form, but new water is not produced. It simply recycles itself.

A stream has nine parts:

> **Head of a Stream:** the starting point of a stream, usually high in the mountains.
>
> **Mouth of a Stream:** the ending point of the stream, it could be into an ocean, river, gulf, etc.
>
> **Channel:** the groove in the landscape through which the stream flows.
>
> **Bed:** the bottom of the channel.
>
> **Banks:** the sides of the channel.
>
> **Flood Plain:** the flat area adjacent to the stream into which a stream overflows (floods).
>
> **Tributaries:** feeder streams that drain the adjacent landscape.
>
> **Drainage Basin:** the region from which a stream drains water. Stream size is usually a function of the drainage basin size.
>
> **Drainage Divide:** an area of positive relief (hill) that separates drainage basins.

Key Factors of a Stream:

Discharge: the volume of water flowing past a given point over a specific length of time.

Gradient (Slope): the perpendicular angle of the stream channel.

    Gradient is extremely important in determining the effects of Streams. It affects the way a stream changes the landscape. Gradient determines what is moved, the clarify of the stream, And the speed of water movement within a stream.

Velocity: how fast a stream is moving.

    Velocity depends on the gradient or slope of a stream, and the steeper the slope, the higher the velocity. The slope is usually steepest at the Head of a stream.

Streams change the landscape by eroding, transporting, and depositing the landscape material.

Groundwater

Ground water is water located beneath the ground's surface. This can be found in the moisture of the soil, but is most commonly thought of as underground areas of trapped water or underground rivers. An underground formation is called an aquifer when it holds a usable quantity of water. Groundwater is part of a perpetual cycle of natural discharge (such as springs and wetlands), and absorption. The depth where soil is completely saturated with water is called the water table.

Skill 21.3   **Analyze atmospheric data (e.g., temperature, barometric pressure, relative humidity, dew point) and how it correlates to weather patterns, climatic conditions, severe weather, and cloud development.**

Humidity: the amount of water vapor in the air. Mass, volume, and density all affect the amount of humidity.

Mass on Earth is equal to weight.

Volume: how much space something takes up. A gaseous mass does not change with a change in volume.

Density: the mass of a substance divided by its volume. Density will change if the volume of a mass changes.

Absolute humidity: the mass of the water vapor divided by the volume of air. The resulting number in calculating absolute humidity is actually meaningless because the mass of water vapor may stay the same as volume changes.

Specific humidity: the mass of the water vapor divided by the mass of the air. This takes into account that the volume of water vapor doesn't change with a change in mass. The formula for specific humidity is useful if you want to know the specific amount of water vapor present, but meteorologically it is not a good indicator. This is because warmer air (air with more energy), holds more water vapor, but it may have the same mass as colder air. In that situation, the colder air would have a lower specific humidity (actually have less water vapor), but mathematically, the warmer air would seem drier.

Relative Humidity: the actual water vapor content divided by the water vapor capacity multiplied by 100. Relative humidity looks at the total amount of water vapor a mass of area could hold (capacity), in relation to how much it actually holds.

As the temperature changes, the capacity of the air to hold water changes: as air cools, water condenses, as air warms, water evaporates. The air is constantly changing temperature as it is heated by the sun or cooled by altitude.

Relative humidity is used in weather prediction and is expressed as a percentage.

100% humidity: The air is saturated with water vapor and can't hold any more without upsetting the equilibrium between water vapor and liquid water.

50% humidity: There is only one-half as much water vapor in the air as it could actually hold, or: At 50% humidity, the air could evaporate two times the water vapor if you increase the air's temperature. In this case, the capacity increases, but the relative humidity actually decreases.

If the temperature goes down, the amount of water vapor present is not altered, but the water vapor capacity decreases and the relative humidity increases.

Another way to look at humidity is by how much evaporation could occur.

If there is 100% relative humidity, then the air is saturated and no more evaporation can occur. But, if there is 50% relative humidity, then evaporation is more likely to occur.

Dew Point: the temperature to which air must be cooled for condensation to occur. If there is a large difference between the dew point and the temperature, then you have low humidity. If there is a small difference between the dew point and the temperature, then you have a higher humidity. Remember: Colder air has a lower water vapor capacity than does warm air. That is why you can have high humidity in cold air and medium humidity in warm air. Ironically, the warm air actually has more water vapor present (specific humidity).

Warm air rises and will continue to rise until it meets air that is as cold or colder than the rising air. When rising air cools to dew point, clouds form due to condensation. Air can rise even if it is only a few degrees warmer than the surrounding air.

Atmospheric stability is determined by comparing the temperature change of an ascending or descending air parcel with the temperature profile of the ambient air layer in which the parcel ascends or descends. Air is considered as either **stable** or **unstable**.

An air layer becomes more stable when it descends, and less stable when it ascends. Unstable air will remain unstable until it encounters air of the same temperature or colder.

In stable air, there is no vertical movement. In unstable air, there is a great deal of vertical movement.

Clouds are classified by their physical appearance and given special Latin names corresponding to the cloud's appearance and the altitude where they occur. Classification by appearance results in three simple categories: cirrus, stratus, and cumulus clouds. Cirrus clouds appear fibrous. Stratus clouds appear layered. Cumulus clouds appear as heaps or puffs, similar to cotton balls in a pile. Classification by altitude result in four groupings: high, middle, low, and clouds that show vertical development. Other adjectives are added to the names of the clouds to show specific characteristics.

Cloud Classifications

High Clouds: -13 °F (-25 °C) Composed almost exclusively of ice crystals.
Cirrus >23,000 ft (7,000 m). Nearly transparent, delicate silky strands (mare's tails), or patches.
Cirrostratus >23,000 ft (7,000 m). A thin veil or sheet that partially or totally covers the sky. Nearly transparent, the sun or moon readily shines through.
Cirrocumulus >23,000 ft (7,000 m). Small, white, rounded patches arranged in a wave or spotted mackerel pattern.

Middle Clouds: 32 - -13 °F (0 - -25 °C) Composed of supercooled water droplets or a mixture of droplets and ice crystals.
Altostratus 6600 - 23,000 ft (2000 - 7000 m). Uniform white or bluish-gray layers that partially or totally obscure the sky layer.
Altocumulus 6600 - 23,000 ft (2000 - 7000 m). Roll-like puffs or patches that form into parallel bands or waves.

Low Clouds: > 23 °F (-5 °C) Composed mostly of water droplets.
  Stratocumulus 0-6,600 ft (0-2000 m). Large irregularly shaped puffs or rolls separated by bands of clear sky.
  Stratus 0-6600 ft (0-2000 m). Uniform gray layer that stretches from horizon to horizon. Drizzle may fall from the cloud.
  Nimbostratus 0-13,120 ft (0-4000 m). Thick, uniform gray layer from which precipitation (significant rain or snow) is falling.

Clouds with Vertical Development: Water droplets build upward and spread laterally.
  Cumulus 0-9840 ft (0-3000 m). Resemble cotton balls dotting the sky.
  Cumulonimbus 0-9840 ft (0-3000 m). Often associated with thunderstorms, these large puffy, clouds have smooth or flattened tops, and can produce heavy rain and thunder.

Skill 21.4   Explain historic and current technologies and tools associated with data collection and interpretation of meteorologic and climatologic research and predictions.

Meteorology: the scientific study of the atmosphere and the atmospheric processes.

Weather Stations

The National Weather Service runs the 500 active weather stations used for national forecasting. However, there are thousands of local weather stations that contribute observations. They collect temperature, precipitation, and humidity data and provide remarks about local conditions.

The data is collected 4 times daily, every 6 hours, around the clock, and sent to the National Weather Bureau. They in turn provide it to the World Meteorological Association, who use it for making long-term predictions, not immediate forecasting.

Meteorological data is collected on the basis of GMT time (Zulu Time). **GMT** refers to Greenwich Mean Time. Greenwich, England is located on the prime meridian, or 0 degrees. Time is measured as either plus or minus the current GMT. Each line of longitude is equal to 4 minutes of time and 15 degrees of longitude is equal to 1 hour of time. You subtract hours as you go west, and add hours as you go east. Example: 8 AM Zulu is equal to 6am at 30° west longitude.

Collection Devices

Weather Balloons: Launched twice a day at 1200 and 2400 Zulu, weather balloons carry a **radiosonde**: an electronic device that measures temperature, pressure, and humidity.

Measurements are taken throughout the ascent of the balloons and this real-time data is radioed back to the surface during the ascent. Weather stations use radars to track the balloons to gather wind speed and direction information.

When the balloon reaches its maximum height, an altimeter releases the radiosonde package via parachute for recovery and reuse. However, only 20% of the radiosondes are recovered. These balloons can also be launched from a plane to gather special data on immediate meteorological disturbances (i.e. hurricanes, storms).

Ground Based Radar: "Doppler Effect" radar is the primary type used because it is very accurate, sensing motion as well as range and direction. This allows for pinpoint accuracy over an area. Example: This is the type of radar you see on television. The **NEXRAD** Radar. (Next Generation Doppler Radar) emits beams of energy that are reflected by the water droplets in the atmosphere. This type of radar is very useful for tracking and predicting rain and less useful for snow or sleet.

Satellites: The National Weather Service heavily depends on its network of weather satellites (i.e. GOES Satellite), to provide a wide-area coverage of the Earth. These satellites primary provide infrared, water vapor, and photographic data and are used to track the formation, development and motion of major meteorological events such as hurricanes and tropical storms.

In terms of orbit, there are two types of satellites: Geostationary and Polar Orbiting. **Geostationary satellites** move with the Earth's rotation. Since they always look at the same point, this allows for a view showing changes over periods of time. **Polar Orbiting satellites** follow an orbit from pole to pole. The Earth rotates underneath the satellite and gives a view of different areas. In effect, it produces slices of the Earth.

The **GOES Satellite** is a key source of weather information. GOES is a geostationary satellite that primarily scans the Atlantic Ocean & U.S. East Coast. There are 4 detection bands on the GOES satellite.

1. Visible Spectrum: Photographic surveys. Pictures are actually light reflecting off the tops of clouds. This capability is limited to daytime use because there is no reflected light during the nighttime hours.

2. Near IR: Looks at heat. Emphasis is on detecting naturally emitted and reflected infrared radiation.

3. Enhanced IR: Adds colors to the tops of the clouds. Uses an enhanced IR (Infrared) receiver/detector to look at reflected or emitted heat. Hotter areas show up as red/yellow/orange, while colder areas appear as blue/green/violet.

4. Water Vapor: Uses radar to detect the water vapor in the atmosphere. The swirls produced by this type of satellite show the air rising. It looks for uplifts that provide an indicator of weather change. The higher the cloud, the colder it is. The rising clouds can indicate thunderstorms, and high, well-developed clouds can signal the advance of a storm formation.

Using the Collected Data

The collected data is sent to the National Meteorological Center in Maryland where it is plotted and then sent on to other weather centers throughout the world. The collected data is used to develop mathematical models that are used to predict the weather. These models have been used since the late 1950's and place differing emphasis on the various aspects of the equational factors involved. A typical model will consist of 6-8 different equational factors.

In turn, the data elements are disseminated to the National Weather Service, which uses them to calculate model results based on 12, 24, 36, 48, and 72-hour predictions. However, the further out you extend the predictions, the less accurate the prediction is likely to be. This is because of the wide variation in the data elements that can occur between the time of prediction and the realization of the prediction time.

The data model is also disseminated to local meteorological offices and radio and television stations for local interpretation. The resulting local forecast often includes a great deal of **Kentucky Windage**: experience based guesses by local meteorologists, and can vary considerably from the forecasts of the National Weather Service. The accuracy of local forecasting is dependent on the experience and expertise of the local meteorologist. Forecasting is only as good as the data collected and local intuition (experience) modifications. Four main types of forecast are developed:

1. Persistence: This doesn't really change much and is very inaccurate.

2. Steady State (Trend): This is a guesstimate as to what the system will do based on the assumption that there will be no change in the data parameters.

3. Analog: This is the best type of forecast basis. It takes the data numbers and plugs them into equations, equations into models, and uses the models for forecasting. This is also called Numerical Forecasting.

4. Climatological: This type of forecast is often vague and is based on long-standing trends. Example: The weather predictions in the "Old Farmer's Almanac."

COMPETENCY 22.0 UNDERSTAND AND APPLY KNOWLEDGE OF INTERACTIONS BETWEEN HUMANS AND EARTH SYSTEMS.

Skill 22.1 Identify the types and characteristics of Earth's renewable and nonrenewable resources and the methods employed to preserve or conserve them.

The two categories of natural resources are renewable and nonrenewable. Renewable resources are unlimited because they can be replaced as they are used. Examples of renewable resources are oxygen, wood, fresh water, and biomass. Nonrenewable resources are present in finite amounts or are used faster than they can be replaced in nature. Examples of nonrenewable resources are petroleum, coal, and natural gas.

Strategies for the management of renewable resources focus on balancing the immediate demand for resources with long-term sustainability. In addition, renewable resource management attempts to optimize the quality of the resources. For example, scientists may attempt to manage the amount of timber harvested from a forest, balancing the human need for wood with the future viability of the forest as a source of wood. Scientists attempt to increase timber production by fertilizing, manipulating trees genetically, and managing pests and density. Similar strategies exist for the management and optimization of water sources, air quality, and other plants and animals.

The main concerns in nonrenewable resource management are conservation, allocation, and environmental mitigation. Policy makers, corporations, and governments must determine how to use and distribute scare resources. Decision makers must balance the immediate demand for resources with the need for resources in the future. This determination is often the cause of conflict and disagreement. Finally, scientists attempt to minimize and mitigate the environmental damage caused by resource extraction. Scientists devise methods of harvesting and using resources that do not unnecessarily impact the environment. After the extraction of resources from a location, scientists devise plans and methods to restore the environment to as close to its original state as possible.

Skill 22.2 Recognize the effects of human activities (e.g., development of urban areas, use of fossil fuels, production of ozone-depleting compounds on the whole earth (e.g., global climate, water pollution, species destruction, erosion).

Humans have a tremendous impact on the world's natural resources. The world's natural water supplies are affected by human use. Waterways are major sources for recreation and freight transportation. Oil and wastes from boats and cargo ships pollute the aquatic environment. The aquatic plant and animal life is affected by this contamination. To obtain drinking water, contaminants such as parasites, pollutants and bacteria are removed from raw water through a purification process involving various screening, conditioning and chlorination steps. Most uses of water resources, such as drinking and crop irrigation, require fresh water. Only 2.5% of water on Earth is fresh water, and more than two thirds of this fresh water is frozen in glaciers and polar ice caps. Consequently, in many parts of the world, water use greatly exceeds supply. This problem is expected to increase in the future.

Plant resources also make up a large part of the world's natural resources. Plant resources are renewable and can be re-grown and restocked. Plant resources can be used by humans to make clothing, buildings and medicines, and can also be directly consumed. Forestry is the study and management of growing forests. This industry provides the wood that is essential for use as construction timber or paper. Cotton is a common plant found on farms of the Southern United States. Cotton is used to produce fabric for clothing, sheets, furniture, etc. Another example of a plant resource that is not directly consumed is straw, which is harvested for use in plant growth and farm animal care. The list of plants grown to provide food for the people of the world is extensive. Major crops include corn, potatoes, wheat, sugar, barley, peas, beans, beets, flax, lentils, sunflowers, soybeans, canola, and rice. These crops may have alternate uses as well. For example, corn is used to manufacture cornstarch, ethanol fuel, high fructose corn syrup, ink, biodegradable plastics, chemicals used in cosmetics and pharmaceuticals, adhesives, and paper products.

Other resources used by humans are known as "non-renewable" resources. Such resources, including fossil fuels, cannot be re-made and do not naturally reform at a rate that could sustain human use. Non-renewable resources are therefore depleted and not restored. Presently, non-renewable resources provide the main source of energy for humans. Common fossil fuels used by humans are coal, petroleum and natural gas, which all form from the remains of dead plants and animals through natural processes after millions of years. Because of their high carbon content, when burnt these substances generate high amounts of energy as well as carbon dioxide, which is released back into the atmosphere increasing global warming. To create electricity, energy from the burning of fossil fuels is harnessed to power a rotary engine called a turbine.

Implementation of the use of fossil fuels as an energy source provided for large-scale industrial development.

Mineral resources are concentrations of naturally occurring inorganic elements and compounds located in the Earth's crust that are extracted through mining for human use. Minerals have a definite chemical composition and are stable over a range of temperatures and pressures. Construction and manufacturing rely heavily on metals and industrial mineral resources. These metals may include iron, bronze, lead, zinc, nickel, copper, tin, etc. Other industrial minerals are divided into two categories: bulk rocks and ore minerals. Bulk rocks, including limestone, clay, shale and sandstone, are used as aggregate in construction, in ceramics or in concrete. Common ore minerals include calcite, barite and gypsum. Energy from some minerals can be utilized to produce electricity fuel and industrial materials. Mineral resources are also used as fertilizers and pesticides in the industrial context.

Skill 22.3   Analyze how the use of natural resources affects human society economically, socially, and environmentally

Biological diversity is the extraordinary variety of living things and ecological communities interacting with each other throughout the world. Maintaining biological diversity is important for many reasons. First, we derive many consumer products used by humans from living organisms in nature. Second, the stability and habitability of the environment depends on the varied contributions of many different organisms. Finally, the cultural traditions of human populations depend on the diversity of the natural-world

Many pharmacological products of importance to human health have their origins in nature. For example, scientists first harvested aspirin, a derivative of salicylic acid from the bark of willow trees. In addition, nature is also the source of many medicines including antibiotics, anti-malarial drugs, and cancer fighting compounds. However, scientists have yet to study the potential medicinal properties of many plant species, including the majority of rain forest plants. Thus, losing such plants to extinction may result in the loss of promising treatments for human diseases.

The basic stability of ecosystems depends on the interaction and contributions of a wide variety of species. For example, all living organisms require nitrogen to live. Only a select few species of microorganisms can convert atmospheric nitrogen into a form that is usable by most other organisms (nitrogen fixation). Thus, humans and all other organisms depend on the existence of the nitrogen-fixing microbes. In addition, the cycling of carbon, oxygen, and water depends on the contributions of many different types of plants, animals, and microorganisms. Finally, the existence and functioning of a diverse range of species creates healthy, stable ecosystems. Stable ecosystems are more adaptable and less susceptible to extreme events like floods and droughts.

Aside from its scientific value, biological diversity greatly affects human culture and cultural diversity. Human life and culture is tied to natural resources. For example, the availability of certain types of fish defines the culture of many coastal human populations. The disappearance of fish populations because of environmental disruptions changes the entire way of life of a group of people. The loss of cultural diversity, like the loss of biological diversity, diminishes the very fabric of the world population.

Skill 22.4 Recognize the effects of Earth processes on human societies through time (e.g., flooding, earthquakes, volcanic eruptions).

Earth processes have a profound effect on human societies. Such effects include choices of where people live, loss of life, technologies that have been developed in response, and the long term effects of natural disasters.

Location

Earth processes have a great deal to do with where people live. The minerals from volcanic eruptions form very fertile soil. Therefore, many villages and farming communities are often found at the base of volcanoes to reap the benefits of the fertile soil. Floodplains have also dictated where people have lived. Although many cities have been located along rivers for transportation purposes, it was not until recently that people began to build more and more on floodplains.

Technologies developed

Humans have developed many technologies in order to survive earth's natural processes. An example of this is the development of earthquake safe structures. In some earthquake prone areas, humans have developed architecture to reverberate or counter the seismic waves. By forming this technology, many structures have had only slight damage in major earthquakes and remain standing. An active earthquake safe building has layers of foam core, rubber, and steel beneath the foundation. These layers absorb the shock of the earthquake. A seismograph in the building reads the frequency of the earthquake waves and pistons in the joints of the building are moved to counteract the tremors. Technology has also been developed to protect against floods. In many areas of the world, levies and pumping systems have been developed in order to protect people from rising water.

Loss of Life

The most obvious effect that natural processes have had on human societies is through the great loss of life that natural processes can cause. One of the most famous of these natural disasters, but not the most destructive, is the loss of life at Pompeii due to the explosion of Mt. Vesuvius in 79 AD. People were caught in pyroclastic flows and were cast in stone. These casts have been excavated and serve as a reminder of earth's destructive force.

Skill 22.5 Analyze the use, quality, and scarcity of global freshwater resources, including protection and conservation measures.

Global freshwater supply

The total water supply on earth is 326 million cubic miles. Each cubic mile is equal to over 1 trillion gallons of water. Only 2.5% of the earth's water is freshwater. Of the freshwater on the earth, 68.7% is locked up in ice caps and glaciers. Groundwater makes up 30.1% of the freshwater. Merely 0.266% of freshwater is found as fresh surface water. The remaining freshwater is found in swamps, the atmosphere and as biological water.

Use of freshwater supply

Freshwater use in the United States is equal to 408 billion gallons of water per day.[1] Among uses of freshwater supply in the United States is domestic use, mining, industrial use, and irrigation. Please view chart below.

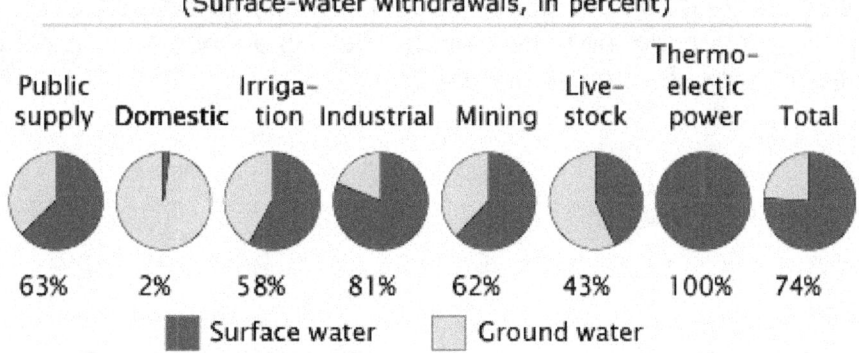

Figure: http://ga.water.usgs.gov

---

[1] http://ga.water.usgs.gov

Use of the freshwater supply around the world has become an important issue. Many areas, including in the United States, face grave water shortages that have significant impact on the land, plant life, and people. The global use of freshwater supply is doubling every 20 years, which is more than twice the rate of population growth.[2] Many countries are facing water stress or water scarcity issues. Globally, more than 1 billion people lack access to freshwater supply.[2] If current trends continue, over 90% of all freshwater resources will be in use by 2025.[2]

Water quality

Global freshwater supplies are being marred by natural and man-made pollutants. There are two types of contamination of water supply: point source pollution and non-point source pollution. Point source pollution can be traced to a specific point of discharge, such as a manufacturing plant. Non-point source pollution cannot be traced to a single point. This type of pollution has its source in runoff over chemically treated land, sewage overflow, and eroded soil. Threats to water quality include microorganisms, chemical pollutants, and nitrogen and phosphorus from animal waste. In some areas, water quality has become extremely poor, causing illness and death.

Water protection and conservation measures

In 1977, the United States enacted the Clean Water Act as a response to major concerns in water quality in the United States. The Clean Water Act specifies water quality standards for industry. It also specifies use for specific bodies of water. Conservation of water has also become important. Over the years, conservation measures such as drip irrigation, low flow showerheads, low flow toilets, and water recycling in industry has increased in use. In times of drought, mandatory conservations have been enacted in many communities. By protecting and conserving the water supply, there is an increased likelihood for the future availability of water for our use.

---

[2] http://www.ozh2o.com/h2use.html

SUBAREA V. ASTRONOMY

COMPETENCY 23.0 UNDERSTAND AND APPLY KNOWLEDGE OF THE CHARACTERISTICS AND FORMATION OF THE PLANETS AND SOLAR SYSTEM.

Skill 23.1 Analyze the effects of gravitational force in the solar system.

The mass of any celestial object may be determined by using Newton's laws of motion and his law of gravity.

For example, to determine the mass of the Sun, use the following formula:

$$M = \frac{4\pi^2}{G} = \frac{a^3}{P^2}$$

where M = the mass of the Sun, G = a constant measured in laboratory experiments, a = the distance of a celestial body in orbit around the Sun from the Sun, and P = the period of the body's orbit.

In our solar system, measurable objects range in mass from the largest, the Sun, to the smallest, a near-Earth asteroid. (This does not take into account, objects with a mass less than 10^{21} kg.)

The surface temperature of an object depends largely upon its proximity to the Sun. One exception to this, however, is Venus, which is hotter than Mercury because of its cloud layer that holds heat to the planet's surface. The surface temperatures of the planets range from more than 400 degrees on Mercury and Venus to below -200 degrees on the distant planets.

Most minor bodies in the solar system do not have any atmosphere and, therefore, can easily radiate the heat from the Sun. In the case of any celestial object, whether a side is warm or cold depends upon whether it faces the sun or not and the time of rotation. The longer rotation takes, the colder the side facing away from the sun will become, and vice versa.

If the density of an object is less than 1.5 grams per cc, then the object is almost exclusively made of frozen water, ammonia, carbon dioxide, or methane. If the density is less than 1.0, the object must be made of mostly gas. In our solar system, there is only one object with that low a density -- Saturn. If the density is greater than 3.0 grams per cc, then the object is almost exclusively made of rocks; and if the density exceeds 5.0 grams per cc, then there must be a nickel-iron core. Densities between 1.5 and 3.0 indicate a rocky-ice mixture.

The density of planets correlates with their distance from the Sun. The inner planets (Mercury-Mars) are known as the terrestrial planets because they are rocky, and the outer planets (Jupiter and outward) are known as the icy or Jovian (gaslike) planets.

In order for two bodies to interact gravitationally, they must have significant mass. When two bodies in the solar system interact gravitationally, they orbit about a fixed point (the center of mass of the two bodies). This point lies on an imaginary line between the bodies, joining them such that the distances to each body multiplied by each body's mass are equal. The orbits of these bodies will vary slightly over time because of the gravitational interactions.

Skill 23.2   Identify the general characteristics (e.g., atmospheric, geological, orbital data) of the sun, the planets, and their satellites.

The solar system is divided into two sections: the inner and outer planets. The inner planets' composition reflects the attraction of the heavier elements by the sun. The outer planets' composition reflects the lighter, less dense elements not attracted as much by the sun's gravitational mass. Heavy elements sink inward to form the core. Lighter elements form the atmosphere.

Our solar system consists of the Sun, planets, comets, meteors, and asteroids. The planets are ordered in the following way: Mercury, Venus, Earth, Mars (with Asteroid Belt), Jupiter, Saturn, Uranus, and Neptune. Pluto has historically been considered the ninth planet. Recently, scientists have created new requirements for the definition of a planet, and it is possible that Pluto will be considered only a celestial body. A common memory aid for remembering the order of the planets is My Very Educated Mother Just Served Us Nine Pies.

All the planets revolve around the sun, and all the planets with the exception of Venus, rotate on their axis in the same direction. Venus has a retrograde motion; it rotates backwards. Except for Pluto, all the planets follow roughly the same elliptical orbital planes around the sun. Neptune and Pluto occasionally change places in the order. Pluto's orbit is very erratic compared to the other planets and sometimes it carriers Pluto inside of Neptune's orbit. The asteroid belt is located between Mars and Jupiter and may be the remnants a planet crushed by the massive gravitational force of Jupiter.

Comparison of the Basic Characteristics of the Inner and Outer Planets:

INNER PLANETS	OUTER PLANETS
Referred to as the Terrestrial Planets	Called the Gas Giants or Jovian Planets
Similar to density to Earth.	Except for Pluto, very large in size.
Also referred to as the "Rocky Planets."	Primarily composed of gas.
Relatively small in size.	Less dense than Earth.
Spin slowly on their axis.	Rotate rapidly on their axis.
Few if any, moons.	Lots of moons.
Mercury	Jupiter
Venus	Saturn (Ringed)
Earth	Uranus
Mars	Neptune
+ Asteroid Belt	Pluto (Rocky)

The Inner Planets:

MERCURY	
Average Distance from the Sun	0.387 AU (5.79×10^7 km)
Density	5.44 g/cm^3
Diameter (equatorial)	4878 km (0.38 of Earth)
Mass	3.31×10^{23} kg (0.055 of Earth)
Atmosphere	None (Trace gasses only of H, He, Na, K)
Axial Rotation Period	58.6 days
Solar Revolution Period	87.9 days
Axial Tilt	0°
Surface Temp	-173°C to 330°C
Moons	None
Composition	Silicate & Iron rocks

VENUS	
Average Distance from the Sun	0.723 AU (1.082 x 10^8 km)
Density	5.3 g/cm^3
Diameter (equatorial)	12,104 km (0.95 of Earth)
Mass	4.87 x 10^{24} kg (0.82 of Earth)
Atmosphere	Runaway Greenhouse effect 96.5% CO_2, 3.5% N.
Axial Rotation Period	243.01 days
Solar Revolution Period	224.68 days
Axial Tilt	177° (Makes the planet appear as if it was rotating clockwise)
Surface Temp	472°C
Moons	None
Composition	Unknown, crustal materials believed to be similar to Earth

Within our solar system, the standard unit of distance measurement is the AU (Astronomical Unit). The AU is the mean distance between the Sun to the Earth. 1 AU = 1.495979 x 10^{11}m.

EARTH	
Average Distance from the Sun	1.00 AU (1.495 x 10^8 km)
Density	5.497 g/cm^3
Diameter (equatorial)	12,756 km
Mass	5.976 x 10^{24} kg
Atmosphere	78% Nitrogen, 21% Oxygen
Axial Rotation Period	24.00 hours
Solar Revolution Period	365.26 days
Axial Tilt	23.5°
Surface Temp	-50°C to 50°C
Moons	1 The Moon

MARS	
Average Distance from the Sun	1.523 AU (2.279 x 10^8 km)
Density	3.94 g/cm^3
Diameter (equatorial)	6796 km (0.53 of Earth)
Mass	0.6424 x 10^{24} kg (0.1075 of Earth)
Atmosphere	95% CO_2
Axial Rotation Period	24.61 hours
Solar Revolution Period	686.9 days
Axial Tilt	23° 59'
Surface Temp	-140°C to 20°C
Moons	2 (*Deimos* and *Phobos*) (Panic and Fear). They may actually be captured asteroids because of irregular shape.
Composition	Identical to Earth's

The Asteroid Belt:

Located between Mars and Jupiter, the asteroid belt is composed of over 2,000 rocky bodies circling the Sun. The largest of these asteroids is over 1,000 km but the smallest are mere particles. The composition of the asteroids is similar to meteorites that have struck the Earth.

The Outer Planets:

JUPITER	
Average Distance from the Sun	5.202 AU (7.783 x 10^8 km)
Density	1.34 g/cm^3
Diameter (equatorial)	142,900 km (11.20 of Earth)
Mass	1.899 x 10^{27} kg (317.83 of Earth)
Atmosphere	Hydrogen, Helium, and Ammonia
Axial Rotation Period	9.83 hours
Solar Revolution Period	11.867 years
Axial Tilt	3.5°
Surface Temp	29,727 °C at surface -120°C in atmosphere
Moons	47, of which only 4 are significant. These are the Galilean Moons: *Io, Europa, Ganymede, and Callisto*.
Composition	No surface, gaseous atmosphere

SATURN	
Average Distance from the Sun	9.538 AU (14.27 x 10^8 km)
Density	0.69 g/cm^3
Diameter (equatorial)	120,660 km (9.42 of Earth)
Mass	5.69 x 10^{26} kg (95.17 of Earth)
Atmosphere	Hydrogen and Helium
Axial Rotation Period	10.65 hours
Solar Revolution Period	29.461 years
Axial Tilt	26° 24'
Surface Temp	Unknown at surface -180° in atmosphere
Moons	17 total of which *Titan* is the largest and most significant.
Composition	No surface, gaseous atmosphere

URANUS	
Average Distance from the Sun	9.538 AU (14.27 x 10^8 km)
Density	1.29 g/cm^3
Diameter (equatorial)	51,118 km (4.01 of Earth)
Mass	8.69 x 10^{25} kg (14.54 of Earth)
Atmosphere	Hydrogen, Helium, and Methane
Axial Rotation Period	17.23 hours
Solar Revolution Period	84.013 years
Axial Tilt	97° 55'
Surface Temp	Unknown at surface -220° in atmosphere
Moons	15 total of which *Oberon* is the largest at 1,500 km diameter.
Composition	Unknown, frozen gasses

NEPTUNE	
Average Distance from the Sun	30.061 AU (44.971 x 10^8 km)
Density	1.66 g/cm³
Diameter (equatorial)	49,500 km (3.88 of Earth)
Mass	10.30 x 10^{26} kg (17.23 of Earth)
Atmosphere	Hydrogen, Helium, and Methane
Axial Rotation Period	16.05 hours
Solar Revolution Period	164.793 years
Axial Tilt	28° 48'
Surface Temp	Unknown at surface -216°C in atmosphere
Moons	8 total, 2 of significance, *Triton* and *Nereid*.
Composition	Unknown, mostly ice

PLUTO	
Average Distance from the Sun	39.44 AU (59.00 x 10^8 km)
Density	2.0 g/cm³
Diameter (equatorial)	2,300 km (0.19 of Earth)
Mass	1.2 x 10^{22} kg (0.002 of Earth)
Atmosphere	Nitrogen and Methane
Axial Rotation Period	9.3 days
Solar Revolution Period	247.7 years
Axial Tilt	122°
Surface Temp	-230°C
Moons	1 *Charon*
Composition	Frozen nitrogen & rock

Skill 23.3   Identify the characteristics and orbital nature of comets, asteroids, and meteoroids.

Comets: small icy bodies that orbit the Sun and produce a glowing tail of gas and dust as the comet nears the Sun. Comets are not planets.

Asteroids: small, rocky worlds that orbit the Sun. Asteroids are sometimes called the minor planets.

Meteoroids: a meteor in space. Meteoroids often are asteroid fragments, but they do not have a specific orbit. Meteoroids can become either meteors, or meteorites depending on whether they burn up or make it through the atmosphere.

Meteors: small fragments of rock debris that turn incandescent and burn up upon entering the Earth's atmosphere.

Meteorites: meteor fragments that do not burn up in the atmosphere and actually strike a planet.

Kepler's Law of Planetary Motion:

Johannes Kepler (1600 A.D.): Danish astronomer Johannes Kepler postulated what is now known as Kepler's Laws of Planetary Motion.

Kepler's First Law of Planetary Motion: A planet can't travel in a circle. A planet travels an elliptical path, with the Sun at one foci point.

Ellipse: a geometric figure drawn around two points, called foci.

Kepler's Second Law of Planetary Motion: Planets must sweep out equal areas, at equal amounts of time, on these elliptical paths. When a planet is nearer to the Sun on the planets elliptical path, it must move faster to sweep over the same area.

Kepler's Third Law of Planetary Motion: A planet's orbital period squared is proportional to its average distance from the sun cubed ($a^3 = p^2$).

Example: Jupiter's average distance to the Sun is 5.20 AU. If a equals 5.20 AU, then a^3 equals 140.6 AU.

Therefore, the orbital period of Jupiter must be the square root of 140.6, which equals a period of about 11.8 light-years.

The significance of Kepler's Law is that it overthrew the ancient concept of uniform circular motion, which was a major support for the geocentric arguments. Kepler suggested that the eccentricity of the planetary motion could be mathematically summarized as $e=c/a$.

The two major points on the elliptical orbit are the Perihelion and Aphelion:

Perihelion: the closest point of approach to the Sun on the elliptical path. This is the point at which the planet travels the fastest in keeping with Kepler's 2nd Law.

Aphelion: the farthest point away from the Sun on the elliptical path. This is the point at which the planet travels the slowest in keeping with Kepler's 2nd Law.

Skill 23.4 **Recognize the scientific basis for understanding various atmospheric, solar, and celestial phenomena (e.g., eclipses, seasons, phases of the moon, relative and apparent motions of objects), meteors, and auroras.**

Seasonal change on Earth is caused by the orbit and axial tilt of the planet in relation to the Sun's Ecliptic: the rotational path of the Sun. These factors combine to vary the degree of insolation at a particular location and thereby change the seasons.

Equinox and Solstice

There are four key points on the Ecliptic. The Equinoxes and the Solstices.

Winter Solstice (December 21) = Shortest day of the year in the northern hemisphere.

Summer Solstice (June 21) = Longest day of the year in the northern hemisphere.

Vernal Equinox (March 21) = Marks beginning of spring.

Autumnal Equinox (Sept 21) = Marks beginning of autumn (Fall)

These dates will vary slightly in relation to leap years.

During the summer solstice, insolation is at a maximum in the northern hemisphere, and at a minimum in the southern hemisphere.

Because of the tilt and curvature of the Earth, in order to get the sun directly overhead, you must be between 23.5°N Latitude and 23.5°S Latitude. This is between the Tropic of Cancer and the Tropic of Capricorn.

Another result is that during the summer months in the northern hemisphere, the far northern latitudes receive 24 hours of daylight. This situation is reversed during the winter months, when you have 24 hours of darkness.

Phases of the moon- from Earth, the lighted half of the moon can only be seen when the moon is opposite the sun. The moon is totally dark when it is located in the direction of the sun. The appearances of full dark and full light are divided into four groups or quarters, these quarters are designated as: (1) new moon, (2) first quarter, (3) full moon, and (4) third quarter.

Eclipses - are defined as the passing of one object into the shadow of another object. A **Lunar eclipse** occurs when the moon travels through the shadow of the earth. A **Solar eclipse** occurs when the moon positions itself between the sun and earth.

Skill 23.5 Recognize historical models of the solar system and the historical development of understanding about the solar system.

How and when the universe formed has puzzled humans for millennia. Although many theories have been offered, no one knows for

1. Creationist Theory

The earliest recorded instances of possible explanation all had mystical-religious foundations, which is not surprising, given that the early civilizations had a deep belief in gods.

Just as civilizations differ according to culture, so do the explanations of how, when, and why the universe was created. However, a sense of balance was a common thread of the explanations.

By whatever names ascribed (i.e. good versus evil, Ying and Yang, etc.), balance provided a sense of order and symmetry out of the inexplicable chaos of what was visible in the world.

Most of the civilizations found comfort in a **Creationist Theory**: the belief that a powerful deity or deities created and controlled all things. However, at various stages of development, most religions largely abandoned polytheism, the belief in more than one god, in exchange for a single, all-powerful God. This was particularly true in the western civilizations as Christianity eventually displaced paganism (naturalism) and animism as the dominant system of belief.

2. Early Models of the Universe

The early models of the universe were primarily products of philosophical musings as combined with religious and social tenets, and observable phenomenon.

Geocentric Model

Geocentric Model: the original model of universal object arrangement in which the universe was Earth centered and all other stellar objects rotated around the Earth. The Sun followed a track around the outer edge of the rotational sphere. The ancient Greeks later modified this model into the Greek Celestial Model (Aristotelian Model).

Greek Celestial Model (Aristotelian Model):
The ancient Greek model was also geocentric. The Earth is still at the center of the universe but the planets, Sun, and stars all are on different spheres orbiting around the Earth. God (heaven) is on the outermost sphere. This model was widely accepted because it was based on the philosophical teachings of Aristotle.

Ptolemaic Model:
The Alexandrian Greek Astronomer Ptolemy, observing the planetary motions in 140 A.D., realized that reality didn't fit the Greek model well. He modified the Greek Celestial model to fit his observations while still retaining a geocentric orientation. Ptolemy's model had the moon orbiting around the Sun in an Epicyclic motion. An **epicycle is a retrograde**: backward motion, in which and object spins or moves clockwise.

Early Dissension to the Geocentric

Aristarchus (260 B.C): 400 Years prior to Ptolemy, the Greek Mathematician Aristarchus proposed the first recorded reference to a Heliocentric: sun centered, universe. He based his conclusion on the study of perfect geometry. His calculations showed that the motions of the observable stellar objects could not be possible if they were geocentric. However, his postulation of heliocentricity was vehemently suppressed by the astronomers of the day. The conclusion would have meant that if the Earth weren't at the center, then it would make the Earth less perfect. This ran completely counter to the deeply entrenched and accepted philosophical thought about God, Earth, and the Heavens.

The Heliocentric Model

Nicolaus Copernicus (1507 A.D.)- Based solely on his extensive planetary observations, Polish monk Nicolaus Copernicus, in his paper *"Revolutions of the Heavenly Spheres,"* triggered *an* earth-shaking revision of human thought by publicly proposing that the universe is heliocentric, not geocentric. He also claimed that the Earth's axis was tilted in relation to the Sun, using retrograde motion, only in a different way.

Copernicus knew that his theory would be highly controversial and out of fear of excommunication by church authorities, he didn't publish his paper until near the end of his life. As he expected, his paper raised a storm of argument between scholars and the church authorities. The debate between Copernicus' supporters and detractors grew so heated as to lead to bloodshed on several occasions. Clearly, this new concept was not to be forgotten as was in the case of Aristarchus.

Tycho Brahe (1572 A.D.)- Spurred on by the growing interest in science, Danish nobleman Tycho Brahe rejected a career as a lawyer and politician, and devoted his life to astronomy and mathematics. In 1572, Tycho, as he is usually referred as, observed a brilliant new star in the sky and like other classically trained astronomers of the day, he was puzzled by its appearance. This new star did not fit into the Aristolian scheme of the heavens, which held that the starry regions were perfectly complete. This meant that the star must be between the moon and the Earth. However, by measuring the parallax of motion, his calculations showed that the star must be further away, challenging the Ptolemaic concepts.

His conclusions were published in a small book *"De Stella Nova* (the New Star)" in 1573 and attracted the attention of the King of Denmark, who offered him funds to build a planetary observatory on the Danish isle of Hveen. Tyco's greatest contribution to astronomy was not theoretical. Instead, his true value to the science lay in his observations.

For over 20 years Tycho constructed new and highly improved existing instruments and studied the motion of over 700 stellar objects, collecting the most the most complete data sets of astronomical observations since the early Egyptian times. Amazingly he did this with the naked eye, as the telescope wasn't invented until almost a hundred years after his death. Tycho hired several mathematicians and astronomers to assist him in collecting and compiling the data. Among these individuals was Johannes Kepler, who would continue Brahe's work following Tycho's death in 1610.

Johannes Kepler (1600 A.D.)- In collating Tycho Brahe's data sets, Johannes Kepler discovered that the data wouldn't fully support the Copernican model. In an effort to try and make the data fit, he postulated what is now known as Kepler's Laws of Planetary Motion.

The significance of Kepler's Laws is that it overthrew the ancient concept of uniform circular motion, which was a major support for the geocentric arguments. Although Kepler postulated three laws of planetary motion, he was never able to explain *why* the planets move along their elliptical orbits, only that they did.

Galileo Galilei (1633 A.D.)- One of the truly brilliant minds of his time, Galileo Galilei was an Italian astronomer and inventor who built his first telescope in 1609. Contrary to popular myth, Galileo didn't invent the telescope but used existing plans and gradually improved upon the design.

The major significance of Galileo's observations is that they totally disproved the geocentric model of the universe. In doing so, Galileo totally destroyed the Aristotelian vision of the universe that had been the basis of astronomical and philosophical though for over 2,000 years.

Galileo's destruction of the Aristotelian theory of the universe created a large problem. If Aristotle was correct, what was the truth? People began to openly question blind belief in religious explanations and sought the answer to their questions through science.

3. Scientific Models of the Universe

As civilizations continued to advance so did our understanding of the physical processes affecting the universe. The early scientific models still blended elements of philosophy and religious thought with measurable data, but as our technological base improved, models reflected less and less of those influences to the point that today, it's a commonly accepted practice to totally discount them in modeling the universe.

COMPETENCY 24.0 UNDERSTAND AND APPLY KNOWLEDGE OF STELLAR EVOLUTION, CHARACTERISTICS OF STARS AND GALAXIES, AND THE FORMATION OF THE UNIVERSE.

Skill 24.1 Recognize characteristics of various star types, including all the stages and processes of stellar evolution.

Stars are not all alike. Their energy outputs vary from 111,000th of the energy to 100,000 times the energy, of Earth's sun. The laws of physics tell us that the more energy an object has, the hotter it is. They also relate color to temperature. Therefore, by observing the color of a star, we get information about its temperature.

The Hertzsprung-Russell Diagram

In 1913 American astronomer Henry Norris Russell and Danish astronomer Ejnar Hertzsprung theorized that the energy emitted by a star is directly related to the star's color. Their supporting graph is now known as the **Hertzsprung-Russell Diagram (H-R Diagram):** a graph that shows the relationship between a star's color, temperature, and mass.

On a H-R Diagram the majority (90%) of the stars plotted form a diagonal line called the **Main Sequence**: the region of the H-R Diagram running from top left to bottom right. Hot, blue stars are at the top left; cooler, red stars are at the bottom right. The middle of main sequence contains yellow stars, like the Sun.

Star mass is also shown on the H-R diagram, increasing from the bottom to the top of the main sequence.

The stars at the bottom right have masses about one-tenth of that of Earth's sun, and the masses increase until you reach the top left where there are stars with masses ten times greater than that of the Sun.

The truly large stars are called **Supergiant Stars**: exceptionally massive and luminous stars 10 to 1,000 times brighter than the Sun. and the smallest stars are called **Dwarf Stars**: dying stars that have collapsed in size. Although small in size, dwarf stars are extremely dense. Although Supergiants are extremely large, they may be less dense than the Earth's outer atmosphere.

We can also use the H-R Diagram to illustrate the life cycle of stars.

The Life Cycle of Stars

Stars form in **Planetary Nebulae**: cold clouds of dust and gas within a galaxy, and go through different stages of development in a specific sequence. This theory of star development is called the **Condensation Theory**.

Sequence of Development

In the initial stage, the diffuse area of the nebula begins to shrink under the influence of its own weak gravity. The cloud-like spheres condense into a knot of gasses called a **Protostar**. The original diameter of the protostar is many times greater than the diameter of our solar system, but gravitational forces cause it to continue to contract. This compression raises the internal temperature of the protostar.

When the protostar reaches a temperature of around 10 million °C (18 million °F), nuclear fusion starts, which stops the contraction of the protostar and changes its status to a star.

Nuclear Fusion: the process in which hydrogen atoms fuse together to form helium atoms, releasing massive amounts of energy during the fusion. It's the fusion of atoms, not combustion, which causes the star to shine.

A star's life cycle depends on its initial mass. Red stars have a small mass. Yellow stars have a medium mass. Blue stars have a large mass. Large mass stars consume their hydrogen at a faster rate and have a short life cycle in comparison to small mass stars that consume their hydrogen at a much slower rate. All stars eventually convert a large percentage of their hydrogen to heavier atoms and begin to die. However, just as their mass determines the length of their life, it also determines the pattern they follow in the last stages of their existence.

Lower Main Sequence Stars:

When small and medium mass stars (such as the Sun) consume all of their hydrogen, their inner cores begin to cool. The stars begin to consume the heavier elements produced fusion (carbon and oxygen) and the star's shell tremendously expands outward, causing the star to become a **Giant Star**: large, cool extremely luminous stars 10 to 100 times the diameter of the Sun. Example: In roughly 4.6 billion years from now our Sun will become a giant star. As it expands, its outer layers will reach halfway to Venus.

The dying Giant gives off thermal pulses approximately every 200,000 years, throwing off concentric shells of light gasses enriched with heavy elements. As it enters its last phases of the life cycle its depleted inner core begins to contract, and the Giant becomes a **White Dwarf Star**: a small, slowly cooling, extremely dense star, no larger than 10,000 km in diameter.

The final phase of a lower main sequence star life cycle can take two paths: most main sequence white dwarfs after a few billion years completely burn out to become **Black Dwarfs**: cold, dead stars. However, if a White Dwarf is part of a **Binary Star:** two suns in the same solar system, instead of slowly cooling to become a Black Dwarf, it may capture hydrogen from its companion star. If this happens, the temperature of the White Dwarf soars and when it reaches approximately 10 million °C, a nuclear explosion occurs, creating a **Nova:** a sudden brightening of a lower main sequence star to approximately 10,000 times its normal luminosity; caused by the explosion of the star. A nova reaches its maximum brightness in a short time (one or two days) and then gradually dims as the gasses and cosmic dust cool.

Upper Main Sequence Stars:

The initial sequence of the high mass, upper main sequence stars is identical to the lower mass stars, Planetary Nebulae to Protostar. However, if the protostar accretes enough material, it forms as a **Blue Star**. When a Blue Star has consumed all of its hydrogen it, too, expands outward, but on a much larger scale then experienced by a lower mass star. It becomes a **Supergiant Star**: an exceptionally bright star, 10 to 1,000 times the diameter of the Sun.

The Supergiant's now depleted core cannot support such a vast weight and collapses inward, causing its temperature to soar. When it reaches roughly 599 million °C, it implodes and then explodes, creating a **Supernova**: the massive explosion of an upper main sequence Supergiant star caused by the detonation of carbon within the star.

A supernova releases more energy than Earth's sun will produce in its entire life cycle. The luminosity of a supernova is as bright as 500 million Suns. For example: Chinese astronomers in 1054 recorded the sudden appearance of a new star in what is now known as the Taurus Constellation. Bright enough to be seen during daytime for over a month, it remained visible for over 2 years.

The explosive release of energy in a supernova is so great (1,028 megatons of TNT) as to literally blow the atomic nuclei of the carbon to bits. The shattered mass is accelerated outward at nearly the speed of light (300,000 km/sec. or 186,000 mph).

90 percent of the shattered mass scatters into space, becoming planetary nebulae from which the life cycle may begin anew. The other 10 percent, the core of the star, is blown inward, becoming a **Neutron Star**: a very small - 10 km diameter-core of a collapsed Supergiant star that rotates at a high speed (60,000 rpm) and has a strong magnetic field (1012 Gauss).

A neutron star may capture gas from space, a companion star, or a nearby star and become a **Pulsar**: a neutron star that emits a sweeping beam of ionized-gas radiation. As the pulsar rotates, the beams of light sweep into space similar to a beacon from a lighthouse. Since first discovered in 1967, over 350 pulsars have been catalogued.

The alternate product of a supernova is a **Black Hole**: a volume of space from which all forms of radiation cannot escape. Black Holes are created when a Supergiant star with a mass roughly 3 times that of the Sun implodes.
The inner core of the star is compacted by the supernova into a Singularity: an object of zero radius and infinite density.

A singularity is difficult to picture. Zero radiuses imply objects with size less than an electron, but also possessing a density that precludes the escape of all radiation including light. Although a singularity has yet to be detected, theoretically, they exist in and cause the effects exhibited by Black Holes.

Constellation: a region of the night sky in which a group of stellar objects form a discernible pattern; usually named after mythological gods, animals, objects, or people.

Skill 24.2 Analyze evidence regarding the chemical composition and physical characteristics of stellar and galactic objects.

The planets are divided into two groups based on distance from the sun. The inner planets include Mercury, Venus, Earth, and Mars. The outer planets include Jupiter, Saturn, Uranus, and Neptune.

Planets

Mercury - the closest planet to the sun. Its surface has craters and rocks. The atmosphere is composed of hydrogen, helium and sodium.

Venus - has a slow rotation when compared to Earth. Venus and Uranus rotate in opposite directions from the other planets. This opposite rotation is called retrograde rotation. The surface of Venus is not visible due to the extensive cloud cover. The atmosphere is composed mostly of carbon dioxide. Sulfuric acid droplets in the dense cloud cover give Venus a yellow appearance. Venus has a greater greenhouse effect than observed on Earth. The dense clouds combined with carbon dioxide trap heat.

Earth - considered a water planet with 70% of its surface covered by water. Gravity holds the masses of water in place. The different temperatures observed on earth allow for the different states (solid. Liquid, gas) of water to exist. The atmosphere is composed mainly of oxygen and nitrogen. Earth is the only planet that is known to support life.

Mars - the surface of Mars contains numerous craters, active and extinct volcanoes, ridges, and valleys with extremely deep fractures. Iron oxide found in the dusty soil makes the surface seem rust colored and the skies seem pink in color. The atmosphere is composed of carbon dioxide, nitrogen, argon, oxygen and water vapor. Mars has polar regions with ice caps composed of water. Mars has two satellites.

Jupiter - largest planet in the solar system. Jupiter has 16 moons. The atmosphere is composed of hydrogen, helium, methane and ammonia. There are white colored bands of clouds indicating rising gas and dark colored bands of clouds indicating descending gases. The gas movement is caused by heat resulting from the energy of Jupiter's core. Jupiter has a Great Red Spot that is thought to be a hurricane type cloud.

Saturn - the second largest planet in the solar system. Saturn has rings of ice, rock, and dust particles circling it. Saturn's atmosphere is composed of hydrogen, helium, methane, and ammonia. Saturn has 20 plus satellites.

Uranus - the second largest planet in the solar system with retrograde revolution. Uranus is a gaseous planet. It has 10 dark rings and 15 satellites. Its atmosphere is composed of hydrogen, helium, and methane.

Neptune - another gaseous planet with an atmosphere consisting of hydrogen, helium, and methane. Neptune has 3 rings and 2 satellites.

Pluto - once considered the smallest planet in the solar system; it's status as a planet is being reconsidered . Pluto's atmosphere probably contains methane, ammonia, and frozen water. Pluto has 1 satellite. Pluto revolves around the sun every 250 years.

A vast collection of stars are defined as **galaxies**. Galaxies are classified as irregular, elliptical, and spiral. An irregular galaxy has no real structured appearance; most are in their early stages of life. An elliptical galaxy consists of smooth ellipses, containing little dust and gas, but composed of millions or trillion stars. Spiral galaxies are disk-shaped and have extending arms that rotate around its dense center. Earth's galaxy is found in the Milky Way and it is a spiral galaxy.

Skill 24.3 Evaluate the characteristics and supporting evidence for the various theories of cosmology and cosmogony.

Cosmogony is any theory concerning the coming into existence or origin of the universe, or an origin belief about how reality came to be. The word comes from the Greek word for "cosmos," meaning "order" or "the world", and the root of "to be born, come about". In the specialized context of space science and astronomy, the term refers to theories of creation of the Solar System and their study (for example, the Solar Nebula theory).

Cosmology, from the Greek words "cosmos" meaning "order" and "logos" meaning "logic" or "science" is the study of the order of the universe. It aims to look at the universe in its totality, and therefore, humanity's place in it. This study has a long history involving science, philosophy, esotericism and religion.

Cosmogony, which is more concerned with the origin of the world, can be distinguished from cosmology, which studies the universe at large and throughout its existence. In practice, there is a scientific distinction between cosmological and cosmogonical ideas. Physical cosmology is the science that attempts to explain all observations relevant to the development and characteristics of the universe as a whole. Questions regarding why the universe behaves in such a way have been described by physicists and cosmologists as being extra-scientific, though speculations are made from a variety of perspectives which include extrapolation of scientific theories to untested regimes and philosophical or religious ideas.

In recent times, physics and astrophysics have come to play a central role in shaping what is now known as physical cosmology; in other words, the understanding of the Universe through scientific observation and experiment. This discipline is generally understood to begin with the big bang, an expansion of space from which the Universe itself is thought to have erupted ~13.7 ± 0.2 billion (10^9) years ago. From its violent beginnings and until its very end, scientists then propose that the entire history of the Universe has been an orderly progression governed by physical laws.

In between the domains of religion and science, stands the philosophical perspective of metaphysical cosmology. This ancient field of study seeks to draw intuitive conclusions about the nature of the Universe, man, god and/or their relationships based on the extension of some set of presumed facts borrowed from spiritual experience and/or observation. Views about the creation (cosmogony) and destruction (eschatology) of the Universe play a central role in shaping a framework of religious cosmology for understanding humanity's role in the Universe.

Skill 24.4 Recognize the historical progression of astronomy and the physical laws that govern it.

Sir Isaac Netwon (1687 A.D.): Today considered the "Father of Modern Physics," Newton is justifiably famous for the development of Calculus and his work with the Laws of Motion.

However, he also turned his brilliant intellect to the phenomena of gravity, postulating what has become known as Newton's Law of Universal Gravitation.

Newton's Law of Universal Gravitation:

Newton's Law of Universal Gravitation: every object attracts every other object with a force that for any two objects is directly proportional to the mass of each object. In simpler terms, objects with mass attract one another. Newton had some difficulty with his own theory; light appeared to be attracted by gravity, but light has no mass. Despite Newton's continued efforts he was unable to solve this paradox, and the problem went unresolved for almost three hundred years.

Albert Einstein (1916 A.D.): Arguably one of the true geniuses of all time, Albert Einstein resolved the Newtonian problem of light appearing to be affected gravitational force.

Einstein's Theory of Gravity:

Einstein compared space to a giant sheet. He theorized that space in not actually flat, but is curved or warped by massive objects.

Einstein's theory describes a Black Hole: an object with infinite with infinite mass and infinite space. The gravity of such a body is to a point that nothing ever escapes, including light.

Einstein showed that the apparent phenomenon of light being attracted by gravity isn't valid. The phenomenon actually represents the presence of another massive body that is blocking the light by warping (curving) the space around it.

COMPETENCY 25.0 UNDERSTAND AND APPLY KNOWLEDGE OF THE HISTORY AND METHODS OF ASTRONOMY AND SPACE EXPLORATION.

Skill 25.1 Demonstrate knowledge of the relationship between latitude and the apparent position and motion of celestial objects.

The positions of the stars are referenced in relation to the Earth and are described in terms of the celestial coordinates of Right Ascension and Declination.

The declination and right ascension is described in reference to the **Celestial Sphere/Globe Model.** You can picture this model as having the Earth centered in the middle of a sphere. The outer framework of the sphere rotates clockwise (east to west) in relation to the Earth. The north and south poles of the Earth correspond to the north and south celestial poles. The Earth's equator corresponds to the celestial equator. **Right Ascension** is roughly analogous to lines of Longitude and is measured eastward from the vernal equinox. **Declination** is roughly analogous to lines of Latitude and is measured in relation to the celestial equator: positive to the north, and negative to the south.

Right Ascension and Declination are measured in units of degrees and time. 1 sec = 1/3600 of a degree. Example: 15° 12' 5" Right Ascension. The sky shifts 15° every hour. That's why photo-telescopes must move with the stars.

Skill 25.2 Describe the various technologies used to observe and explore space (e.g., various types of telescopes, deep-space probes, artificial satellites) and the scientific laws that govern space flight.

There are two main styles of telescopes: refractor and reflector.

Refractor Telescopes

Refraction: the bending of light. Example: Put a straw in a clear glass of water. Now look at straw through the side of the glass. It will appear to bend at the point where the straw enters the water.

Refractor Telescope: an optical device that makes use of lenses to magnify and display received images. Professional astronomers do not use Refractor telescopes because of two main problems: first the telescopes are affected by chromatic aberrations which make it difficult to focus on the stars, and second, because they rely solely on lenses, the telescopes have inherent weight and size restrictions. Chromatic Aberrations are a problem because as the lenses split the light into its chromatic components; each color has a different focal point based on its wavelength. This causes details on the image to blur.

To compensate for this problem lenses with different refractive indexes (the degree of bending) are sandwiched together in order to compensate for the aberrations and properly focus on the image. However, this compensation method increases the size and weight of the telescope.

Reflector Telescopes

Reflection: the re-emission of light off of an object struck by the light. Example: Look at yourself in the mirror. What your eyes see is the re-emission of light waves that have struck you and the mirror.

Reflector Telescope: an optical device that makes us of a mirror or mirrors, to reflect light waves to an eyepiece (an ocular), thereby eliminating chromatic aberrations. There are different types of reflector telescopes; some use mirrors only, and others make use of both lenses and mirrors.
The objective is the primary focus mirror in a reflector telescope and is usually curved. Any other mirrors in the telescope are referred to as secondary, tertiary, etc. depending on the number of mirrors present. On both refractor and reflector telescopes, the eyepiece is called the ocular. The most common style of reflector telescope used is the Schmidt-Cassegrain. The two major advantages of a reflector telescope over a refractor style telescope are:

- No chromatic aberrations results in a much clearer and detailed image than in a refractor telescope.

- Their smaller size makes them easier to use, and they weigh less because they don't rely on the heavy, thick glass lenses required in a refractor type.

However, there are still limits to the size of a Reflecting telescope. They are very expensive and it is an extremely difficult and slow process to grind the mirror to the proper specifications. The larger the mirror or mirrors required to view distant objects, the more likely it is that there will be **imperfections** in the mirror. Weight also becomes an issue depending on the size of the mirror required. The heavier and thicker the mirror required, the greater the chance that the mirror will **sag,** slightly distorting the image. Also, the mirror will heat and cool unevenly, causing further distortion.

Active Optics

Active Optics: a type of optical device that is composed of hexagonal pieces of mirror whose positions are controlled by a computer. Also referred to as **Active Telescopes**.

Collectively, smaller pieces of mirror weigh less than a single large mirror and, more importantly, they generally do not suffer from sagging problems. Small hexagonal shaped pieces of mirror are arranged next to each other to form a larger reflection surface. Computer-controlled thrusters mounted underneath the pieces control the mirror position and focus. The use of the smaller pieces positioned so as to work in conjunction, eliminates sagging and the uneven heating and cooling problems found in extremely large, single mirror type telescopes.

Example: The Keck Telescope in Hawaii is able to have a 10-meter diameter reflective surface through the use of active optics. Similarly, the *Hobby-Eberly* Telescope composed of 91 hexagons has an 11-meter diameter reflective surface, making it the largest telescope on Earth.

New Generation Telescopes

Hubble Space Telescope: Named in honor of American astronomer Edwin Hubble, who proved the theory of an expanding universe. The **Hubble Space Telescope,** although only possessing a 2.4 meter diameter reflective surface, isn't affected by atmospheric constraints, and as a result, it provides a much clearer, higher resolution image of stellar objects than is possible through the use of a Earth-based telescope. Two companies competed in the design phase of manufacturing the telescope, and working models of each design were constructed. However, shortly after being positioned in space, the mirror on the winning design was discovered to have imperfections that produced a great deal of distortion in the received images. Initially, the fix for this problem seemed simple; put the losing contractor's design up in space in place of the flawed telescope. But a problem quickly arose. When NASA scientists contacted the second company, they were told that the losing model no longer was available. In any event, NASA eventually figured out a means to correct the imperfections and space shuttle astronauts successfully affected the repairs.

CCD-Charged Coupled Devices

A **CCD** is a camera plate made up of thousands of tiny pixels. The pixels carry a slight electrical charge and when photons strike a pixel, electrons are released. The release of the electrons causes a flow of current through an attached wire, and this current is detected by a computer chip and used to construct images based on the number of strikes. The number of strikes also shows the intensity of the received image.

CCD cameras have a wide range of applications besides astronomy. This type of technology is being successfully employed in many of the newest, high-resolution cameras available to the general public.

Radio Telescopes

Variances in radio waves received from space can be translated into usable astronomical data. The advantages offered by use of radio telescopes are many: it's cheaper to build a radio telescope than optical telescopes, they can operate 24 hrs a day and be built just about anywhere on Earth, and they open up an entirely new window of space investigation. But they initially had one major disadvantage: the useable radio waves received from space were not overly abundant and generally very weak, and you needed a huge receiving dish to detect the signals. To overcome these problems, scientists developed a technique called Radio Interferometry.

Radio Interferometry: a method of amplifying weak radio waves by lying out in a Y-shaped pattern, a series of small radio telescopes all pointed at the same point in the sky. The telescopes add their received signals together to form an overlay of signals. Computers control the angle of incidence and correlate the incoming signals. This improves resolution and limits the size of the radio telescope dish needed for a single unit.

Skill 25.3   Recognize the scope and scale of astronomical time and distance.

Measurement Units in Astronomy

Astronomical distances represent mind-boggling amounts of distance. Because our standard units of distance measurement (i.e. kilometers) would result in so large of a number as to become almost incomprehensible, physicists use different units of measurement to reference the vast distances involved in astronomy.

Within our solar system, the standard unit of distance measurement is the **AU (Astronomical Unit).** The **AU** is the mean distance between the Sun to the Earth. 1 AU = 1.495979×10^{11} m. Outside of our solar system, the standard unit of distance measurement is the **Parsec.** 1 Parsec = 206,265 AU or 3.26 Light Years (LY).

Light year (LY): the distance light travels in one year. As the speed of light is 3.00×10^8 m/sec, one light year represents a distance of 9.5×10^{12} km, or 63,000 AU.

Measuring the Distance to the Stars

The distances are measured using a shift in viewpoint. This is called the **Parallax**: an apparent change in the position of an object due to a change in the location of the observer. In astronomy, parallax is measured in seconds of arc. 1 second of arc = 1/3600 of a degree.

The concept of measuring distances by parallax is based on the mathematical discipline of trigonometry. Example: Photos of the distant stars are taken at different times, usually 6 months apart. The apparent shift in position that the star has moved in comparison to the previous photo is the parallax.

By measuring the angles, we can use trigonometric functions of sine, cosine, and tangent to determine a distance to the object. The smaller the parallax is, the greater the distance to the star.

The Components and Properties of Light

The Components of Light

Light is composed of both visible and invisible aspects. For example, a rainbow is visible while radio waves are invisible.

What Light Looks Like

Light is actually a waveform that travels by electromagnetic fields. Unlike sound, light can travel through a vacuum. Sound travels due to an increase or decrease of particle vibrations through the air. Sound does not travel in a vacuum.

We measure light in wavelengths. The symbol for wavelength is the Greek letter (λ) Lambda. The distance between the start of and the end of one full cycle of the waveform is called the wavelength.
The distance between the top and midpoint of a waveform is called the amplitude.

The Properties of Light

The speed of light is: 3×10^8 m/sec. This is expressed by use of the symbol *"c"*. The speed of the waveform is expressed in the units of Hertz (Hz). This is a function of wavelength times the frequency (*f*). **Frequency**: the number of complete waveform cycles that pass a fixed point in one second (1 Hz = 1/*f*). In other words, $c = 2f$. Wavelength and frequency have an inverse relationship. If frequency goes up, the wavelength goes down. If wavelength goes up, the frequency goes down.

Skill 25.4 Recognize the historical technologies used to determine distance and time and their impact on civilization and progress.

Measuring distance

In 1400 BC, the Greek Herodotus wanted to survey land for the estimate of taxes. He had surveyors tie knots in ropes at set intervals. Then "rope stretchers" attached plumb bobs and used sighting instruments to form accurate measurements.

Eventually the Greeks abandoned the ropes, and used a triangulation method and built instruments in order to measure the angles. The Romans used 10 foot wooden rods with metal ends and laid them end to end in order to measure distances.

Edmund Gunter invented a survey chain in 1620 that was comprised of 100 iron links measuring 66 ft. long. This allowed him to measure more accurately uneven areas (such as a property line along a stream).

The first official survey of the east coast of the United States was done by Ferdinand Hassler. He standardized iron bars that were not as affected by temperature and humidity and clamped them together to be exactly 8 meters long. These would be laid down to measure distances. These rods were cumbersome and slow to use, and "invar" was created as a nickel steel hybrid. The invar was used to make metal tapes that did not have the expansion and contraction problems of steel.

In 1948, it was discovered that light could be used to measure distance. In the 1970's Doppler was used to measure distance. Finally, GPS and satellites are currently used with extreme accuracy. Through proper measurement, accurate maps were created and property lines were delineated.

Measuring time

Prehistoric man used changes in seasons, stars, and day and night to measure time. This helped prehistoric man plan growing seasons and nomadic activity.

The sundial was one of the first formal technologies for measuring time. The sundial is a round disk that has a wedge on it. Around the circle are time markings. As the sun rises in the sky, the wedge casts shadows around the dial, measuring the time against the markings.

The hourglass and the clepsydra (water clock) worked in similar ways. They both worked by allowing sand or water drip through a narrow opening to measure time.

Early clocks measured time by falling weights or springs. Then, pendulums were used to tick off time. Finally, electronic motors and the vibration of quartz when a current is passed through are used to measure the passing of time. Measuring time allowed civilizations to form calendars, plan religious ceremonies, and plan farming and hunting times of the year.

Skill 25.5   Describe the historic progression of space exploration and the technologies that made it possible.

One of the primary means of learning more about the planetary bodies of the solar system is through space exploration. Beginning in the 1960's when the first probes journeyed toward Earth's Moon; a planned sequence of spacecraft has visited some of the planetary objects in our solar system.

Mercury:

Mercury was visited in the early 1970's.

Mariner 10 probe:

Launched in November 1973, the probe reached Mercury in 1974. It orbited the planet three times, collecting basic photographic data. Since the initial visit, no other probes have been sent there.

Venus:

Venus has had lots of exploration, visited primarily during the 1960's and 1970's. Twelve probes have been sent there.

Mariner 4, 5, and 10. (Conducted Orbital recon.)

Venera 4, 7, 12, 15, and 16. (Soviet Union). (Mostly orbital recon.)

Venera 7 actually landed on the surface.

Pioneer (Orbital recon.)

Magellan (Orbital recon.) The most recent and most accurate data.) Venus's surface is very hot! Electronics have trouble surviving the heat and this limits attempts to land.

Mars:

Visited in the late 1960's, early 1970's, and then again in the 1990's.

Viking 1 and 2: Viking 1 actually landed on the surface. It was the first probe on Mars.

Pathfinder & Sojourner: Sojourner actually probed the surface with a small remote controlled robotic all terrain vehicle probe.

The Outer Planets:

Pioneer 10 and 11 visited Jupiter in the early 1970's.

Voyager 1 and 2:
>Voyager 1: Visited Jupiter and Titan (a Moon of Saturn.) (Orbital recon.)
>Voyager 2: Visited Jupiter, Saturn, Uranus, Neptune, and then exited the solar system. Voyager 2 is the only probe to visit all 4 of the Jovian planets.

Recent Explorations

Galileo: Launched in 1989, Galileo is currently orbiting Jupiter.
Cassini-Huygens: Launched in 1997, it was scheduled to reach Saturn in 2004. Cassini is the spacecraft. Huygens is the probe to be sent down to Titan.

Titan was chosen because it has an atmosphere and other probes have detected possible water phases. Additionally, Titan's density indicates it's more Earth-like than gaseous Saturn is. Given the possibility of water and a solid surface, conditions may exist to support some form of life.

Stardust: Launched in 1998, it is currently orbiting the Sun for the second time. After it finishes the 2nd solar orbit, it will go to a comet and will collect material from the tail of the comet for return to Earth. This marks the first time (except for the moon) that collected physical specimens will be returned to Earth.

TEACHER CERTIFICATION STUDY GUIDE

Sample Test

Directions: The following are multiple choice questions. Select from each grouping the best answer.

1. Which layer of the atmosphere would you expect most weather to occur?

A. troposphere
B. thermosphere
C. mesosphere

2. What percentage of earth's surface is covered by water?

A. 61%
B. 71%
C. 81%

3. Which layer of the earth's atmosphere contains the Ozone layer?

A. thermosphere
B. troposphere
C. stratosphere

4. Copernicus developed a theory that is known as

A. baycenter
B. heliocentric
C. geocentric

5. The boundary that separates the crust from the mantle is known as

A. Moho
B. shadow zone
C. catacastic

6. The product of intrusive activities would result in forming a

A. cinder cone
B. volcanic pipe
C. dike

7. A star's light and heat are produced by

A. magnetism
B. electricity
C. nuclear fusion

8. The center of an atom is called

A. micron
B. nucleus
C. electron

9. An important food source for animals in a water biome

A. shrimp
B. plankton
C. seaweed

10. The smallest piece of an element is called a/an

A. compound
B. nucleus
C. atom

11. An instrument that measures relative humidity is known as

A. psychrometer
B. anemometer
C. barometer

EARTH & SPACE SCIENCE

12. Removing salts from ocean water by heating is called

A. filtration
B. distillation
C. freezing

13. When molecules in the air cool and combine to form rain, _____ has occurred.

A. condensation
B. convection
C. radiation

14. Which instrument measures wind direction?

A. anemometer
B. barometer
C. wind vane

15. The most important cause of erosion is

A. water
B. wind
C. air

16. An anemometer measures

A. wind velocity
B. temperature
C. relative humidity

17. The boundary that develops when a cold air mass meets a warm air mass

A. cold front
B. warm front
C. stationary front

18. When the sun, moon and earth are aligned in a straight line what type of tides are produced?

A. neap tides
B. high tides
C. spring tides

19. North of the equator, currents move in which direction?

A. counter-clock wise
B. clockwise
C. northerly

20. Rocks that serve as aquifers are

A. impermeable
B. permeable
C. igneous

21. Volcanoes with violent eruptions are known as

A. shield volcanoes
B. dome volcanoes
C. cinder volcanoes

22. The Richter scale measures

A. compressions
B. focus
C. magnitude

23. The earth's outer core is probably

A. liquid
B. solid
C. rock-bed

24. What are the two most abundant elements found in the earth's crust?

A. oxygen and oxides
B. oxygen and carbonates
C. oxygen and silicon

25. The San Andreas Fault is classified as a

A. transform fault
B. oblique-slip fault
C. reverse fault

26. Batholiths are the largest structures of which type of rock activity?

A. intrusive rock
B. extrusive rock
C. magma

27. The main agents of chemical weathering are

A. water, oxygen, CO_2
B. water, oxygen, sulfur
C. water, oxygen, nitrogen

28. Soil classified as porous is called

A. clay soil
B. laterites soil
C. sandy soil

29. Intrusive igneous rock forms

A. glassy texture
B. small crystals
C. large crystals

30. Which types of rocks are rich sources of fossil remains'?

A. sedimentary rock
B. metamorphic rock
C. intrusive rock

31. The best preserved animal remains have been discovered in

A. resin
B. lava
C. tar-pits

32. The Mid-Atlantic is a major area of which type plate movement?

A. subduction plate movement
B. divergent plate
C. convergent plate

33. When lava cools quickly on the earth's surface the newly formed rock is called

A. clastic
B. intrusive
C. extrusive

34. When a dyke forms with magma flowing in a tub-like structure this is known to be a/an

A. extrusive activity
B. intrusive activity
C. metaphoric activity

35. A caldera is formed when a large depression collapses. This is the result of a

A. sinkhole
B. aquifer
C. volcanic eruption

36. Trenches observed on the sea floor are the results of

A. interaction
B. divergence
C. subduction

37. Alfred Wegener's hypothesis of continental drift was not supported until scientists began studying

A. sea floor
B. mountain ranges
C. volcanoes

38. These massive waves are caused by the displacement of ocean water, and are often the result of underwater earthquakes.

A. epicenters
B. tidal waves
C. tsunamis

39. Sea floor spreading occurs when the earth's crust is stretched and pulled apart in a process called

A. slippage
B. rifting
C. drifting

40. The layer of the atmosphere that is in a plasma state and aids in communication

A. Thermosphere
B. Ionosphere
C. Mesosphere

41. A stream erodes bedrock by grinding sand and rock fragments against each other. This process is defined as:

A. dissolving
B. transportation
C. abrasion

42. Rocks formed from magma are:

A. igneous
B. metamorphic
C. sedimentary

43. Rocks formed by the intense heating or compression of pre-existing rocks are classified as

A. igneous
B. metamorphic
C. sedimentary

44. Rocks made of loose materials that have been cemented together are:

A. igneous
B. metamorphic
C. sedimentary

45. River valley glaciers produce

A. U-shaped erosion
B. V-shaped erosion
C. S-shaped erosion

46. The life cycle of a river with the most cutting power and erosion is known as which stage?

A. youth stage
B. mature stage
C. old age

47. The result of radioactive decay

A. parent element
B. daughter element
C. half-life

48. The most abundant dry gas found in the atmosphere is

A. oxygen
B. nitrogen
C. CO_2

49. A natural groundwater outlet through which boiling water and steam explodes into the air is called a _____.

A. sinkhole
B. artesian system
C. geyser

50. Which of the following rocks make the best aquifer?

A. granite
B. basalt
C. sandstone

51. Sediments that settle out from rivers are called

A. deposits
B. boulders
C. sandstone

52. A hole that remains in the ground after a block of glacier ice melts is called a

A. pothole
B. sinkhole
C. kettle

53. The first sign that a tsunami is approaching a shore is

A. a sudden flatten of the waves
B. water moving from the shore
C. large wall of water on horizon

54. Mountains that have been squeezed into wavelike patterns are called

A. fold mountains
B. dome mountains
C. fault-block mountains

55. The largest ocean is the

A. Atlantic
B. Pacific
C. Indian

56. The major surface current that flows along the east coast of the United States is known as the

A. Bermuda Current
B. Mexican Current
C. Gulf Stream

57. The formation of ocean waves is caused by

A. earth's rotation
B. the moon
C. the wind

58. The most abundant compound found in sea water is

A. chloride
B. calcium carbonate
C. magnesium chloride

59. The distance between two meridians is measured in degrees of

A. longitude
B. latitude
C. magnitude

60. A contour line that has tiny comb-like lines along the inner edge indicates a

A. depression
B. mountain
C. valley

61. Fossils that are used to date strata are called

A. datum fossils
B. index fossils
C. true fossils

62. Which of the following causes the aurora borealis?

A. gases escaping from earth
B. particles from the sun
C. particles from the moon

63. The layer of the atmosphere that shields earth from harmful ultraviolet radiation is called

A. ionic layer
B. ozone layer
C. equatorial layer

64. The layer of the earth's atmosphere that is closest to the earth's surface is the

A. stratosphere layer
B. thermosphere layer
C. troposphere layer

65. The sun transfers its heat to other objects by

A. conduction
B. radiation
C. convection

66. As an air mass expands, it becomes

A. cooler
B. warmer
C. denser

67. Air moving northward from the horse latitudes produces a belt of winds called the

A. prevailing westerlies
B. north westerlies
C. trade winds

68. Which type of cloud always produces precipitation?

A. altostratus
B. cirrostratus
C. nimbostratus

69. An air mass that forms over the Gulf of Mexico is called

A. polar
B. maritime
C. continental

70. Spring tides will occur when the moon is in its

A. quarter phases
B. full and new phases
C. half phases

71. Air pressure is measured using a

A. barometer
B. hydrometer
C. physcrometer

72. The two most abundant elements found in stars are

A. hydrogen and calcium
B. hydrogen and helium
C. hydrogen and neon

73. A comet's tail always points _____ from the sun.

A. towards
B. perpendicular
C. away

74. The dark areas observed on the sun are known as

A. solar flares
B. prominences
C. sun spots

75. An example of distance in degrees of latitude is

A. 55° north
B. 93° east
C. 25° west

76. A scale use to measure the hardness of a mineral is known as the

A. Bowen's scale
B. Mohs' scale
C. Harding scale

77. When a gas changes to a liquid this process is known as

A. evaporation
B. condensation
C. dissolution

78. A fan-shaped river deposit is better known as a

A. levee
B. flood plain
C. delta

79. When heat energy is trapped by the gases in the Earth's atmosphere this process is called

A. greenhouse effect
B. coriolis effect
C. constant effect

80. Winds in the Northern Hemisphere are deflected to the

A. north
B. left
C. right

81. Water vapor and _____ trap heat in the atmosphere.

A. carbon dioxide
B. nitrogen
C. sodium nitrate

82. The frontal system that forms when a cold air mass meets a warm air mass and does not change position is defined as a

A. occluded front
B. stationary front
C. warm front

83. Surface ocean currents are caused by which of the following

A. temperature
B. density changes in water
C. wind

84. The length of time it takes for two waves to pass in a row is called

A. wave length
B. wave period
C. wave crest

85. Circulation of the deep ocean currents is the result of

A. equatorial currents
B. surface currents
C. density currents

86. Chains of undersea mountains associated with the spreading of the seafloor are known as _____.

A. ocean trenches
B. mid ocean ridges
C. seamounts

87. A shallow, calm area of water located between a barrier island and a beach area is defined as a/an _____.

A. atoll
B. coral reef
C. lagoon

88. Closed contour lines noticed on a topographical map indicate which type of information?

A. rivers and lakes
B. hills
C. mountains

89. The heliocentric model was developed by which famous scientist?

A. Kepler
B. Copernicus
C. Newton

90. The phases of the moon are the result of its _____ in relation to the sun.

A. revolution
B. rotation
C. position

91. A telescope that collects light by using a concave mirror and can produce small images is called a _____.

A. radioactive telescope
B. reflecting telescope
C. refracting telescope

92. The measuring unit to measure the distance between stars is called

A. astronomical unit
B. light-year
C. parsec

93. The largest planet found in the solar system is

A. Pluto
B. Jupiter
C. Saturn

94. The famous scientist who discovered the elliptical orbits

A. Kepler
B. Copernicus
C. Galilee

95. The planet with retrograde rotation is

A. Pluto
B. Uranus
C. Venus

96. A star's brightness is referred to as

A. magnitude
B. mass
C. apparent magnitude

97. Clouds of gas and dust where new stars originate are called

A. black holes
B. super novas
C. nebulas

98. The transfer of heat from the earth's surface to the atmosphere is called ____.

A. conduction
B. radiation
C. convection

99. The ozone layer is found in the

A. stratosphere layer
B. mesosphere layer
C. exosphere layer

100. The coldest zone of the atmosphere is found in the

A. thermosphere
B. mesosphere
C. stratosphere

101. Winds in high pressure areas tend to blow

A. clockwise
B. counterclockwise
C. along the center

102. When warm air meets cold air this is defined as a

A. cold front
B. occluded front
C. warm front

103. The fastest velocity of a river is found where?

A. bottom
B. center
C. sides

104. As a glacier melts the sea level tends to:

A. rise
B. sink
C. evaporate

105. The largest groups of minerals found in the earth's crust are

A. silicates
B. carbonates
C. quartz

106. Used to measure the magnitude of an earthquake.

A. Richter scale
B. epicometer
C. seismograph

107. These are types of folds:

A. anticlines and synclines
B. faults and folds
C. fractures and shearings

108. Breaks in rocks which indicate movement are known as

A. fractures
B. folds
C. faults

109. The collision of two continental plates is called a

A. folded mountain range
B. volcanic mountain range
C. block mountain range

110. Plates that move in the same direction are termed

A. divergent faults
B. convergent faults
C. transform faults

111. Studying the positions of layered rock is referred to as

A. relative ages
B. index fossils
C. disconformity

112. The smallest division of geologic time is defined as

A. Periods
B. Eras
C. Epochs

113. The most common fossils of the Paleozoic Era are

A. angiosperms
B. trilobites
C. endotherms

114. Contamination may enter groundwater by

A. air pollution
B. leaking septic tanks
C. photochemical processes

115. Which is a form of precipitation?

A. snow
B. frost
C. fog

116. A dead star is called a _____.

A. White Dwarf
B. Super Giant
C. Black Dwarf

117. Roughly ninety percent of all geologic time is said to be _____.

A. Paleozoic
B. Pre-Cambrian
C. Mesozoic

118. The massive change in biological conditions that marked the beginning of life forms on earth is known as _____.

A. Oxygen Revolution
B. Carbon Revolution
C. Trilobite Revolution

119. Water is a truly unique material. It has the property of _____.

A. Adhesion
B. Cohesion
C. Both

120. The following is not a form of satellite used to track weather:

A. NEXRAD
B. Geostationary
C. Polar Orbitting

121. Over the course of our planet's history Earth has had _____ atmosphere(s).

A. one
B. two
C. three

122. Which is not a principle law of geology?

A. Cross Cutting
B. Faulting
C. Super position

123. The red beds are important because they indicate the presence of _____ in the geologic record.

A. Carbon
B. Ammonia
C. Oxygen

124. Tornadoes are most likely to occur in what season?

A. Spring
B. Summer
C. Autumn

125. Which scale is used to measure hurricanes?

A. Fujita Scale
B. Saffir-Simpson Scale
C. Richter Scale

Answer Key

1. A	45. A	89. B
2. B	46. A	90. C
3. C	47. B	91. B
4. B	48. B	92. C
5. A	49. C	93. B
6. C	50. C	94. A
7. C	51. A	95. C
8. B	52. C	96. A
9. B	53. B	97. C
10. C	54. A	98. A
11. A	55. B	99. A
12. B	56. C	100. B
13. A	57. C	101. A
14. C	58. A	102. C
15. A	59. A	103. B
16. A	60. A	104. A
17. A	61. B	105. A
18. C	62. B	106. C
19. B	63. B	107. A
20. B	64. C	108. C
21. C	65. B	109. A
22. C	66. B	110. C
23. A	67. A	111. A
24. C	68. C	112. C
25. A	69. B	113. B
26. A	70. B	114. B
27. A	71. A	115. A
28. C	72. B	116. C
29. C	73. C	117. B
30. A	74. C	118. A
31. C	75. A	119. C
32. B	76. B	120. A
33. C	77. B	121. C
34. B	78. C	122. B
35. C	79. A	123. C
36. C	80. C	124. A
37. A	81. A	125. B
38. C	82. B	
39. B	83. C	
40. B	84. B	
41. C	85. C	
42. A	86. B	
43. B	87. C	
44. C	88. B	

TEACHER CERTIFICATION STUDY GUIDE

Rationales for Sample Questions

1. Which layer of the atmosphere would you expect most weather to occur?
A. Troposphere
The troposphere is the lowest portion of the Earth's atmosphere. It contains the highest amount of water and aerosol. Because it touches the Earth's surface features, friction builds. For all of these reasons, weather is most likely to occur in the Troposphere.

2. What percentage of earth's surface is covered by water?
B. 71%
The earth's surface is nearly ¾ covered with water. The Pacific Ocean is the largest body of moving water. Of course there are other oceans, lakes, rivers, and glaciers as well.

3. Which layer of the earth's atmosphere contains the Ozone layer?
C. Stratosphere
The stratosphere is located above the troposphere and below the mesosphere. It has layers striated by temperature. The warmest portion, the ozone layer, is warm because it absorbs solar ultraviolet radiation.

4. Copernicus developed a theory that is known as
B. Heliocentric
Copernicus' theory stated that the planets revolved around the sun (helios), as opposed to prior belief that the planets revolved around the Earth (geocentric).

5. The boundary that separates the crust from the mantle is known as
A. Moho
The Mohorovicic Discontinuity separates oceanic and/or continental crust from the Earth's mantle.

6. The product of intrusive activities would result in forming a
C. dike
A dike is formed when upwelling magma cools and solidifies beneath the surface, an intrusive activity.

7. A star's light and heat are produced by
C. nuclear fusion
Nuclear fusion is the process in which hydrogen atoms fuse together to form helium atoms, releasing massive amounts of energy during the fusion. It's the fusion of atoms, not combustion, which causes the star to shine.

TEACHER CERTIFICATION STUDY GUIDE

8. The center of an atom is called
B. nucleus
The center of the atom is the nucleus. The nucleus of the atom is composed of nucleons, which when electrically charged are protons and when electrically neutral are neutrons. However, the electrons swirl around the nucleus in a large region called the Electron Cloud.

9. An important food source for animals in a water biome
B. plankton
Drifting organisms that inhabit the water column are called plankton. They may be phytoplankton or zooplankton. Phytoplankton are autotrophs and form the base of the aquatic food chain.

10. The smallest piece of an element is called a/an
C. atom
An atom is the smallest particle of the element that has the properties of that element. All of the atoms of a particular element are the same. The atoms of each element are different from the atoms of the other elements.

11. An instrument that measures relative humidity is known as
A. psychrometer
A psychrometer measures relative humidity. The other choices, anemometer and barometer, measure wind speed and atmospheric pressure, respectfully.

12. Removing salts from ocean water by heating is called
B. distillation
In the process of distilling ocean water the saline water is heated, producing water vapor that is in turn condensed, forming fresh water. The salt is left behind as waste but the water is used in many areas for drinking supply.

13. When molecules in the air cool and combine to form rain, _____ has occurred.
A. condensation
Condensation is the change in matter from a denser phase, such as a gas (or vapor) to a liquid. Condensation commonly occurs when a vapor is cooled to a liquid.

14. Which instrument measures wind direction?
C. wind vane
Of the choices given, an anemometer measures wind speed (velocity), a barometer measures atmospheric pressure and a wind vane indicates wind direction.

TEACHER CERTIFICATION STUDY GUIDE

15. The most important cause of erosion is
A. water
Erosion is most often caused by water. This can be acid rain eroding rocks, rivers eroding riverbeds, oceans eroding beaches and cliffs, etc. In addition, wind is another source of erosion.

16. An anemometer measures
A. wind velocity
Of the choices given, an anemometer measures wind speed (velocity), temperature would be measured by a thermometer, and relative humidity is measured with a psychrometer.

17. The boundary that develops when a cold air mass meets a warm air mass
A. cold front
Fronts are always labeled according to the approaching air mass. Therefore, a cold air mass meeting and displacing a warm air mass would be called a cold front.

18. When the sun, moon and earth are aligned in a straight line what type of tides are produced?
C. spring tides
Spring tides are produced when the Earth, Sun, and Moon are in a line. Therefore, spring tides occur during the full moon and the new moon. Neap tides occur during quarter moons. They occur when the gravitational forces of the Moon and the Sun are perpendicular to one another (with respect to the Earth).

19. North of the equator, currents move in which direction?
B. clockwise
North of the equator, currents move clockwise. South of the equator, currents move counter clockwise.

20. Rocks that serve as aquifers are
B. permeable
Aquifers are underground areas of water-bearing permeable rock from which groundwater can be collected.

21. Volcanoes with violent eruptions are known as
C. cinder volcanoes
Cinder volcanoes are some of the most violent volcanoes because of the immense pressure of gas built up within the neck of the volcanic tube. When it overcomes the resistance offered by the surrounding rock, it rips off the top of the cone. A huge mass of liquid magma and Pyroclastic Rock are flung outward in a violent explosion.

EARTH & SPACE SCIENCE

22. The Richter scale measures
C. magnitude

The richter scale is used to measure the magnitude of earthquakes. Focus and compressions refer to areas of activity, but are not examples of a scale for measuring.

23. The earth's outer core is probably
A. liquid

The earth's inner core is mathematically hypothesized to be a solid iron and nickel core. The outer core, surrounding the inner core, is so hot that it is believed to be molten iron (liquid state). Combined, they are responsible for Earth's magnetism.

24. What are the two most abundant elements found in the earth's crust?
C. oxygen and silicon

Earth's crust is composed of 47% oxygen and 28% silicon.

25. The San Andreas Fault is classified as a
A. transform fault

The San Andreas fault is considered a transform fault because sections of the earth's crust (the Pacific and North American Plates) slide side-by-side past each other.

26. Batholiths are the largest structures of which type of rock activity?
A. intrusive rock

Batholiths are large portions of igneous intrusive rock deep within the Earth's crust that form from cooled magma.

27. The main agents of chemical weathering are
A. water, oxygen, CO_2

Water is the greatest factor in chemical weathering. Glaciers erode entire valleys. Rainfall pounds way at topographic surface features. Rivers erode riverbeds and river edges. Oceans erode shorelines and cliffs. Oxygen is also a factor in weathering. Air movement can have erosional factors. Most importantly, wind can transport material to other areas, having both erosional and depositional results. Carbon dioxide combines with water to produce carbonic acid, which erodes rock structures and some of our man made monuments.

28. Soil classified as porous is called
C. sandy soil

Sandy soil has a high sand content. The sand molecules have many spaces in-between, making the soil porous. This soil does not hold water well.

TEACHER CERTIFICATION STUDY GUIDE

29. Intrusive igneous rock forms
C. large crystals
Intrusive igneous rock forms large crystals. This rock is formed from magma that cools and solidifies within the earth. Because it is surrounded by pre-existing rock, the magma cools slowly, and the rocks are coarse grained. The crystals are usually large enough to be seen by the unaided eye.

30. Which types of rocks are rich sources of fossil remains'?
A. sedimentary rock
Sedimentary rock has the most abundant fossil collection. This is because, over time the layers of sand and mud at the bottom of lakes & oceans turned into rocks due to compression. Plants and animals that died and fell to the bottom were part of the compressional process by which the many layers were eventually turned into stone, encapsulating a fossil.

31. The best preserved animal remains have been discovered in
C. tar pits.
Tar pits provide a wealth of information when it comes to fossils. Tar pits are oozing areas of asphalt, which were so sticky as to trap animals. These animals, without a way out, would die of starvation or be preyed upon. Their bones would remain in the tar pits, and be covered by the continued oozing of asphalt. Because the asphalt deposits were continuously added, the bones were not exposed to much weathering, and we have found some of the most complete and unchanged fossils from these areas, including mammoths and saber toothed cats.

32. The Mid-Atlantic is a major area of which type plate movement?
B. divergent plate
The Mid- Atlantic is home to a submerged mountain range, which extends from the Arctic Ocean to beyond the southern tip of Africa. The divergent plate action results in sea floor spreading at a rate of about 2.5 centimeters per year (cm/yr), or 25 km in a million years, creating the vast ocean we recognize today.

33. When lava cools quickly on the earth's surface the newly formed rock is called
C. extrusive
Rock formed by the cooling of magma on the earth's surface is known as extrusive, as opposed to intrusive, which is formed by the cooling of magma below the Earth's surface.

34. When a dyke forms with magma flowing in a tub-like structure this is known to be a/an
B. intrusive activity
Dykes are thin, vertical veins of igneous rock. They form within fractures in the earth's crust. Intrusive activity forces magma into underground areas, which can seep into these existing fractures forming a dyke.

EARTH & SPACE SCIENCE

35. A caldera is formed when a large depression collapses. This is the result of a
C. volcanic eruption
A caldera is the collapse of land following a volcanic eruption. Once the underground store of magma and gas has been released in a volcanic explosion, there is not enough support, causing the ground to collapse. A caldera is sometimes confused with the area from which magma and gases are emitted (a crater).

36. Trenches observed on the sea floor are the results of
C. subduction
Trenches are created where two plates collide (converge). Plate collision causes denser oceanic crust to sink or slip beneath lighter continental crust. It is subducted and melted into the asthenosphere, producing a deep trench on the ocean floor parallel to the plate boundary.

37. Alfred Wegener's hypothesis of continental drift was not supported until scientists began studying
A. sea floor
Wegener's hypothesis of continental drift was supported by studies of the sea floor. In comparison to continental rock materials, the youngest rock is found on the ocean floor, consistent with the tectonic theory of cyclic spreading and subduction. Overall, oceanic material is roughly 200 million years old, while most continental material is significantly older, with age measured in billions of years.

38. These massive waves are caused by the displacement of ocean water, and are often the result of underwater earthquakes.
C. tsunamis
Earthquakes can trigger an underwater landslide or cause sea floor displacements that in turn, generate deep, omni-directional waves. Far out to sea these waves may be hardly noticeable. However, as they near the shoreline, the shallowing of the sea floor forces the waves upward in a "springing" type of motion.

39. Sea floor spreading occurs when the earth's crust is stretched and pulled apart in a process called
B. rifting
A rift is a place where the Earth's crust and lithosphere are being pulled apart. In rifts, no crust or lithosphere is produced. If rifting continues, eventually a mid-ocean ridge may form.

TEACHER CERTIFICATION STUDY GUIDE

40. The layer of the atmosphere that is in a plasma state and aids in communication
B. Ionosphere
The Ionosphere is an area of free ions: positively charged ions, produced as a result of solar radiation striking the atmosphere. It is known for its production of aurora borealis and its benefits to radio transmission

41. A stream erodes bedrock by grinding sand and rock fragments against each other. This process is defined as
C. abrasion
Abrasion is the key form of mechanical weathering. It is a sandblasting effect caused by particles of sand or sediment. Abrasive agents include wind blown sand, water movement, and the materials in landslides bashing into each other.

42. Rocks formed from magma are
A. igneous
Igneous rocks are rocks that have formed from cooled magma. They are further classified as extrusive or intrusive according to location.

43. Rocks formed by the intense heating or compression of pre-existing rocks are classified as
B. metamorphic
Metamorphism is the process of changing a pre-existing rock into a new rock by heat and or pressure. Metamorphism is similar to that of putting a clay pot into a kiln. The clay doesn't melt, but a solid-state chemical reaction occurs that causes a change. The chemical bonds of adjoining atoms breakdown and allow the atoms to rearrange themselves, producing a substance with new properties.

44. Rocks made of loose materials that have been cemented together
C. sedimentary
Sediments are broken up rock material. Sand on a beach or pebbles in a mountain stream are typical examples. Sedimentary rocks are named for their source; they are rocks that form from sediments that lithify to become solid rock. Sedimentary rock is especially important for the finding of fossils.

45. River valley glaciers produce
A. U-shaped erosion
River valleys are typically V-shaped. The velocity and cutting power of a river is greatest at its center. However, glaciers broaden the area. Upon its retreat, a glacier typically leaves a U-shaped eroded valley.

46. The life cycle of a river with the most cutting power and erosion is known as which stage?
A. youth stage
Young streams have straight paths, no flood plain, a "V" shaped cutting profile, and high velocity with generally clear water and low suspended load. Old streams have lots of meanders, large flood plain, flat profile, low velocity, with murky, "muddy" waters because of a high-suspended load.

47. The result of radioactive decay
B. daughter element
The radioactive decay causes the (mother) element to change into an (daughter) element. The Mother-Daughter relationship of produced nuclides during the series of isotope decay is the basis for radiometric dating. Although many isotopes are used in radiometric dating, the most widely known method is referred to as Carbon-14 dating. Knowing the half-life (how long it takes for half of the material to decay) is the key factor in the radiometric dating process.

48. The most abundant dry gas found in the atmosphere is
B. Nitrogen
The atmosphere is composed of 78% Nitrogen, 21% Oxygen, and 1% other gasses.

49. A natural groundwater outlet through which boiling water and steam explodes into the air is called a
C. geyser
A geyser is a thermal spring that erupts. The processes behind the eruption are very similar to those involved in boiling water in a teakettle. A constriction forms in the connected chambers of a spring. The water heats under pressure, turns to steam, and erupts with great force past the constriction. The ejected steam condenses and returns to a liquid state. The water draws back into its chambers and the process begins again. Since it takes awhile for the water to drain back and reheat, geysers often erupt on a determinable schedule.

50. Which of the following rocks make the best aquifer?
C. sandstone
Sandstone makes the best aquifer because of its porosity. It has larger pores than granite or basalt, and is also likely to fracture in a way that is conducive to water movement and collection.

51. Sediments that settle out from rivers are called
A. deposits
Deposits are pieces of matter that settle out of the water and fall to the bottom, or are washed into a collection area, such as a delta. This can be terrestrial matter, biological matter, salts, or larger pebbles and rocks.

52. A hole that remains in the ground after a block of glacier ice melts is called a
C. kettle

As the outwash moves sediment alongside and in the path of a receding glacier, blocks of ice can be buried beneath the sediment. After years of erosion these blocks are uncovered and melt, leaving a shallow depression behind. When these depressions fill, they are known as Kettles, and become scenic lakes.

53. The first sign that a tsunami is approaching a shore is
B. water moving from the shore

The first sign that a tsunami is approaching is usually the retreat of water from the shoreline. When the water returns, it comes fast and washes well past its normal level in both distance and depth, destroying coastal areas and causing many losses.

54. Mountains that have been squeezed into wavelike patterns are called
A. fold mountains

During mountain building or compressional stress, rocks may deform to produce folds. Generally, a series is produced. The up-folds are called anticlines and the down-folds are known as synclines.

55. The largest ocean is the
B. Pacific

The four major oceans (listed in decreasing size) are the Pacific, Atlantic, Indian and Arctic.

56. The major surface current that flows along the east coast of the United States is known as the
C. Gulf Stream

The Gulf Stream begins in the Caribbean and ends in the northern North Atlantic. It is powerful enough to be seen from outer space and is one of the world's most studied current systems. It acts as the east coast boundary current plays an important role in the transfer of heat and salt to the poles.

57. The formation of ocean waves is caused by
C. the wind

Wind is the primary factor in the production of ocean waves. It is the energy and friction of wind action that transfers to the water to create waves.

58. The most abundant compound found in sea water is
A. chloride

Chloride is the compound found most often in sea water. Other compounds commonly found include sodium carbonate, magnesium and potassium compounds, sulfite, bromide, and silicate. NaCl is what we commonly refer to as sea salt. Of the two components, chloride is more readily available in the sea.

59. The distance between two meridians is measured in degrees of
A. longitude
Longitude describes the location of a place on Earth east or west of a line called the Prime Meridian. Longitude is given in degrees ranging from 0° at the Prime Meridian to 180° east or west

60. A contour line that has tiny comb-like lines along the inner edge indicates a
A. depression
Contour lines are shown as closed circles in elevated areas and as lines with miniature perpendicular lined edges where depressions exist. These little lines are called hachure marks.

61. Fossils that are used to date strata are called
B. index fossils
Index fossils are fossils of organisms that were known to be abundant at specific times in Earth's history. Presence of such fossils gives one an idea of what age the surrounding material came from.

62. Which of the following causes the aurora borealis?
A. particles from the sun
Aurora Borealis is a phenomenon caused by particles escaping from the sun. The particles escaping from the sun include a mixture of gases, electrons and protons, and are sent out at a force that scientists call solar wind. Together, we have the Earth's magnetosphere and the solar wind squeezing the magnetosphere and charged particles everywhere in the field. When conditions are right, the build-up of pressure from the solar wind creates an electric voltage that pushes electrons into the ionosphere. Here they collide with gas atoms, causing them to release both light and more electrons.

63. The layer of the atmosphere that shields earth from harmful ultraviolet radiation is called
B. ozone layer
The ozone layer is the part of the Earth's atmosphere that contains high concentrations of ozone (O_3). It is located in the stratosphere and absorbs UV radiation emitted from the sun, making life possible on Earth.

64. The layer of the earth's atmosphere that is closest to the earth's surface is the
C. troposphere layer
The troposphere is the layer of Earth's atmosphere that is the lowest (closest to the surface). It is the densest because it contains almost all the water vapor and aerosol found in the atmosphere. It is easy to conclude, then, that most weather phenomena occur here.

65. The sun transfers its heat to other objects by
B. radiation

Radiation is the process by which energy is transferred in the form of waves or particles. The Sun emits ultraviolet radiation in UVA, UVB, and UVC forms, but because of the ozone layer, most of the ultraviolet radiation that reaches the Earth's surface is UVA.

66. As an air mass expands it becomes
B. warmer

Air expends as heat is applied according to the laws of gasses.

67. Air moving northward from the horse latitudes produces a belt of winds called the
A. prevailing westerlies

The prevailing westerlies are the winds found in the middle latitudes between 30 and 60 degrees latitude. They blow from the high pressure area in the horse latitudes towards the poles.

68. Which type of cloud always produces precipitation?
C. nimbostratus

Nimbostratus clouds are seen as a thick, uniform, gray layer from which precipitation (significant rain or snow) is falling. Of the other choices offered, altostratus clouds appear as uniform white or bluish-gray layers that partially or totally obscure the sky, and cirrostratus are like a thin, nearly transparent, veil or sheet that partially or totally covers the sky. Only nimbostratus guarantees precipitation.

69. An air mass that forms over the Gulf of Mexico is called
B. maritime

Maritime air masses are moist, containing considerable amounts of water vapor, which is ultimately condensed and released as rain or snow. Maritime tropical air originates near the Gulf of Mexico and travels north-east across the warm Atlantic to affect western Europe, as well as north-west across the United States.

70. Spring tides will occur when the moon is in its
B. full and new phases

Spring tides are produced when the Earth, Sun, and Moon are in a line. Therefore, spring tides occur during the full moon and the new moon. Neap tides occur during quarter moons. They occur when the gravitational forces of the Moon and the Sun are perpendicular to one another (with respect to the Earth).

71. Air pressure is measured using a
A. barometer

A psychrometer measures relative humidity. A barometer measures atmospheric pressure. A hydrometer is used to measure the specific gravity of a liquid.

TEACHER CERTIFICATION STUDY GUIDE

72. The two most abundant elements found in stars are
B. hydrogen and helium
Hydrogen and helium are the only elements that occur naturally in our universe. It makes sense, then, that they are present in all areas, including stars.

73. A comet's tail always points _____ from the sun.
C. away
A comet's tail always points away from the sun. The sun's radiation is burning up the ice that makes the comet, and since it is projecting the material outward, the tail seems to be pointing away from the sun. Notice that this question does not use a specific direction (north, south, east, west) because comets move and are subject to the viewer's location and perception.

74. The dark areas observed on the sun are known as
C. sun spots
Larger dark spots called Sunspots appear regularly on the Sun's surface. These spots vary in size from small to 150,000 kilometers in diameter and may last from hours to months. The sunspots also cause solar flares that can accelerate to velocities of 900 km/hr, sending shock waves through the solar atmosphere.

75. An example of distance in degrees of latitude is

A. 55° north

Latitude is measured in degrees away from the equator. The equator marks 0°, and parallel lines moving around the globe are quantified in degrees north or south.

76. A scale use to measure the hardness of a mineral is known as the
B. Moh's scale
The Moh's scale of hardness measures the scratch resistance of minerals. The hardest material is diamond, and the frailest is talc. This means that diamond can scratch any surface, which is not true of less hard materials, such as talc.

77. When a gas changes to a liquid this process is known as
B. condensation
Condensation is the change in matter to a denser phase, such as a gas (or vapor) to a liquid. Condensation can occur when a vapor is cooled to a liquid or when a vapor is compressed.

78. A fan-shaped river deposit is better known as a
C. delta
Flowing water carries the material to the ocean where one of two things happen, the material is deposited on the offshore continental shelf or is carried back inland to the inlets and bays. Over time, the sediment thickly accumulates and may form typical coastal features such as sand bars and deltas.

EARTH & SPACE SCIENCE

79. When heat energy is trapped by the gases in the Earth's atmosphere this process is called
A. greenhouse effect
When greenhouse gases and heat build up, the Earth's surface and atmospheric temperature rises. The current and controversial hypothesis contends that if we cut the amount of rising CO_2 in the atmosphere, then things will cool down.

80. Winds in the Northern Hemisphere are deflected to the
C. right
The Earth is spinning on its rotational axis. Spin is greatest near the equator and least at the poles. The different velocities associated with the spin give rise to an effect on the air known as the Coriolis Force. The idea is that the result of the Coriolis effect is that winds in the north are deflected to the right, and winds in the south are deflected to the west.

81. Water vapor and _____ trap heat in the atmosphere.
A. carbon dioxide
Water vapor and carbon dioxide are both considered greenhouse gases because they can trap heat in the atmosphere. Other sources of greenhouse gasses include rice paddies and ruminant animals, which produce Methane.

82. The frontal system that forms when a cold air mass meets a warm air mass and does not change position is defined as a
B. stationary front
Fronts are the boundaries where one air mass meets another. A stationary front is a boundary between two air masses when neither is strong enough to displace the other.

83. Surface ocean currents are caused by which of the following
C. wind
A current is a large mass of continuously moving oceanic water. Surface ocean currents are mainly wind-driven and occur in all of the world's oceans (example: the Gulf Stream). This is in contrast to deep ocean currents which are driven by changes in density.

84. The length of time it takes for two waves to pass in a row is called
B. wave period
The wave period is the time required for two successive waves to pass. Wave crest is the tallest part of the wave. Wave length is measured from the crest of one wave to the crest of the next.

TEACHER CERTIFICATION STUDY GUIDE

85. Circulation of the deep ocean currents is the result of
C. density currents
Unlike surface currents, deep ocean currents are driven by changes in density. These density differences may be caused by changes in salinity (halocline) or temperature (thermocline). Colder water sinks below warmer waters, causing a river (current) flowing below the warmer waters.

86. Chains of undersea mountains associated with the spreading of the seafloor are known as
B. mid ocean ridges
Mid ocean ranges are underwater mountains formed by plate tectonics. The underwater mountains are all connected, making a single mid-oceanic ridge system that is the longest mountain range in the world. The ridges are active sites with new magma constantly emerging onto the ocean floor and into the crust, resulting in sea floor spreading.

87. A shallow, calm area of water located between a barrier island and a beach area is defined as a/an
C. lagoon
A lagoon is known for its quiet movement of water. A lagoon is a body of shallow salt or brackish water separated from the sea by a shallow or exposed sandbank, coral reef, etc. Non-reef lagoon barriers are formed by wave-action or longshore currents depositing sediments. Because of their gentle atmosphere and brackish water, they are often nurseries for many baby fish and aquatic animals.

88. Closed contour lines noticed on a topographical map indicate which type of information?
B. hills
The rules of contouring dictate that contour lines are closed around hills, basins, or depressions. Because we know that depressions are shown using hachure marks, a closed contour line without such marks represents a hill.

89. The heliocentric model was developed by which famous scientist?
B. Copernicus
Copernicus is recognized for his heliocentric theory. The heliocentric theory postulates that the heavenly bodies rotate around the sun. Prior to his assertions, people believed in the geocentric model that held that all bodies rotated around the Earth. The geocentric model was supported by the church, so Copernicus' ideas were highly controversial.

90. The phases of the moon are the result of its _____ in relation to the sun.
C. position
The moon is visible in varying amounts during its orbit around the earth. One half of the moon's surface is always illuminated by the Sun (appears bright), but the amount observed can vary from full moon to none.

TEACHER CERTIFICATION STUDY GUIDE

91. A telescope that collects light by using a concave mirror and can produce small images is called a
B. reflecting telescope
Reflecting telescopes are commonly used in laboratory settings. Images are produced via the reflection of waves off of a concave mirror. The larger the image produced the more likely it is to be imperfect.

92. The measuring unit to measure the distance between stars is called
C. parsec
Parsecs are the units used to describe the distance between stars. Astronomical units (AU) are used to describe the distances between celestial objects (example The Earth is 1.00 ± 0.02 AU from the Sun). Light years are a unit of length measuring the distance light travels in a vacuum in one year.

93. The largest planet found in the solar system is
B. Jupiter
The planets (in decreasing size) are Jupiter, Saturn (body- not inclusive of rings), Uranus, Neptune, Earth, Venus, Mars, Mercury (Pluto was thought to be the smallest planet, but is no longer classified as a planet).

94. The famous scientist who discovered the elliptical orbits
A. Kepler
The significance of Kepler's Laws is that it overthrew the ancient concept of uniform circular motion, which was a major support for the geocentric arguments. Although Kepler postulated three laws of planetary motion, he was never able to explain *why* the planets move along their elliptical orbits, only that they did.

95. The planet with retrograde rotation is
C. Venus
Venus has an axial tilt of only 3° and a very slow rotation. It spins in the direction opposite of its counterparts (who spin in the same direction as the Sun). Uranus is also tilted and orbits on its side. However, this is thought to be the consequence of an impact that left the previously prograde rotating planet tilted in such a manner.

96. A star's brightness is referred to as
A. magnitude
Magnitude is a measure of a star's brightness. The brighter the object appears, the lower the number value of its magnitude. The apparent magnitude is how bright an observer perceives the object to be. Mass has to do with how much matter can be measured, not brightness.

97. Clouds of gas and dust where new stars originate are called
C. nebulae

Nebulae are where new stars are born. They are large areas of gasses and dust. When the conditions are right, particles combine to form stars.

98. The transfer of heat from the earth's surface to the atmosphere is called
A. conduction

Radiation is the process of warming through rays or waves of energy, such as the Sun warms earth. The Earth returns heat to the atmosphere through conduction. This is the transfer of heat through matter, such that areas of greater heat move to areas of less heat in an attempt to balance temperature.

99. The ozone layer is found in the
A. stratosphere

The stratosphere is home to the ozone layer, which protects Earth from harmful UV radiation.

100. The coldest zone of the atmosphere is found in the
B. mesosphere

The mesosphere is the coldest layer of the atmosphere, with temperatures as low as -100°Celsius. Within this layer, temperature decreases with increasing altitude.

101. Winds in high pressure areas tend to blow
A. clockwise

High pressure systems are known for winds that flow clockwise and fair weather. Low pressure systems are accompanied by clouds and precipitation and winds flow counterclockwise.

102. When warm air meets cold air this is defined as a
C. warm front

When a warm air mass meets and displaces a cold air mass, the front is called a warm front.

103. The fastest velocity of a river is found where?
B. center

Mountain streams have little fining (sorting the material by size) due to their higher velocity, and low land streams are muddy because the velocity is less and erosion occurs on the bed and sides of the stream. Once a stream is at or close to base level, equilibrium is achieved between deposition and erosion. Erosion and deposition are controlled by the velocity of the stream. As the stream approaches base level, more of its energy is in a side-to-side cutting (meanders) than in down-cutting.

TEACHER CERTIFICATION STUDY GUIDE

104. As a glacier melts the sea level tends to
A. rise
As a glacier melts, its water is distributed into nearby bodies of water, causing the sea level to rise.

105. The largest groups of minerals found in the earth's crust are
A. silicates
Silicates are the most abundant group of minerals found in the Earth's crust. The two most abundant elements in the earth's crust are Oxygen (46.6%) and Silicon (27.7%). These combine together to form silicates, which some scientists believe make up as much as 90% of the Earth's crust.

106. Used to measure the magnitude of an earthquake.
C. seismograph
A seismograph is a machine used to measure the magnitude of an earthquake. As the Earth's materials move, the weight also moves and sends an electronic signal to a recording device called a seismograph. Movements are displayed as a series of lines on a recording chart called a Seismogram, reflecting the seismic energy detected at a particular location.

107. These are types of folds:
A. anticlines and synclines
Folded mountains are composed of up and down folds. The up-folds are called anticlines and the down-folds are known as synclines.

108. Breaks in rocks which indicate movement are known as
C. faults
Faults are rock fractures that indicate relative movement. Fractures are also breaks in the rock, but they show no evidence of movement. Folds are created from compression and are a forming of tectonic building.

109. The collision of two continental plates is called a
A. folded mountain range
The collision of two continental plates results in a folded mountain range. Two continental plates pushing against each other but not subducting, will cause the material to buckle, sometimes repeatedly, giving these mountains their characteristic ribbon appearance.

110. Plates that move in the same direction are termed
C. transform faults
Transform faults are areas where two plates move in the same direction. They are parallel, and do not collide, but may result in earthquakes if areas of the plates stick or have excessive pressure in sliding past each other.

TEACHER CERTIFICATION STUDY GUIDE

111. Studying the positions of layered rock is referred to as
A. relative ages
The Earth's materials-rocks, soils, and sediments-are piled upon each other in layers called strata. Understanding the relative orientation and arrangement of the strata provides important information about the Earth's history and the ongoing sequence of events and processes that helped shape that history.

112. The smallest division of geologic time is defined as
C. epochs
Geologic time is divided into eons, eras, periods, and epochs (listed here in decreasing order of size).

113. The most common fossils of the Paleozoic Era are
B. trilobites
Trilobites flourished in the Paleozoic era. There were over 600 genera and 1000's of species. Trilobites were bottom dwellers and scavengers found in shallow to deep water. For an extremely long period of time, Trilobites were the dominant multi-cellular life form on the planet. Trilobites are very good guide fossils because they were extremely abundant and existed throughout the entire Paleozoic period. Their development underwent distinctive changes, and these differences are useful in subdividing the time period.

114. Contamination may enter groundwater by
B. leaking septic tanks
Leaking septic tanks allow contamination to slowly seep into the ground, where it is absorbed into the water table and infects the groundwater.

115. Which is a form of precipitation?
A. snow
Snow is a form of precipitation. Precipitation is the product of the condensation of atmospheric water vapor that falls to the Earth's surface. It occurs when the atmosphere becomes saturated with water vapor and the water condenses and falls out of solution. Frost and fog do not qualify as precipitates.

116. A dead star is called a _____.
C. Black Dwarf
The final phase of a lower main sequence star's life cycle can take two paths: most main sequence white dwarfs after a few billion years completely burn out to become what is called a black dwarf: a cold, dead star. Alternatively, if a White Dwarf is part of a Binary Star: two suns in the same solar system, instead of slowly cooling to become a Black Dwarf, it may capture hydrogen from its companion star.

117. Roughly ninety percent of all geologic time is said to be _____.
B. Pre-Cambrian
Pre-Cambrian Time: Comprised of the Hadean, Archean, and Proterozoic Eons, 87% of all geologic time is considered Pre-Cambrian.

118. The massive change in biological conditions that marked the beginning of life forms on earth is known as _____.
A. Oxygen Revolution
Between 4.6 and 3.6 billion years ago, we transition from an uninhabitable Earth, to the appearance of simple, single-celled bacteria. Around 2.5 billion years ago, the bacteria developed the ability of photosynthesis. This process released oxygen as a by-product and there was a massive release of oxygen as the bacteria multiplied. This massive release is called the Oxygen Revolution and it concurrently marks the beginning of the Proterozoic Eon.

119. Water is a truly unique material. It has the property of _____.
C. Both
A unique property of water is that water likes itself; it has a natural tendency to stick to itself. This property is based upon the polar nature of the water molecule. It attracts other water molecules. When the molecules stick together, they are attached through Hydrogen Bonds, giving the molecule a property called cohesion. Cohesion gives water an unusually strong surface tension, and its capillary action makes the water spread. When the water spreads, adhesion, the tendency of water to stick to other materials, allows water to adhere to solids, making them wet.

120. The following is not a form of satellite used to track weather:
A. NEXRAD
While all of these instruments are used to track weather, the NEXRAD Radar, Next Generation Doppler Radar, is not a satellite. It emits beams of energy that are reflected by the water droplets in the atmosphere. This type of radar is very useful for tracking and predicting rain and less useful for snow or sleet. Geostationary satellites move with the Earth's rotation. Since they always look at the same point, this allows for a view showing changes over periods of time. Polar Orbiting satellites follow an orbit from pole to pole. The Earth rotates underneath the satellite and gives a view of different areas. In effect, it produces slices of the Earth.

121. Over the course of our planet's history Earth has had _____ atmosphere(s).
C. three
Earth's initial atmosphere was composed of primarily hydrogen and smaller amounts of helium. However, most of the hydrogen and helium escaped into space very shortly after the earth was formed, approximately 4.6 billion years ago. A second atmosphere formed during the first 500 million years of Earth's history, as the gasses trapped within the planet were out- gassed during volcanic eruptions. This atmosphere was composed of carbon dioxide (CO_2), Nitrogen (N), and water vapor (H_2O), with smaller amounts of methane (CH_4), ammonia (NH_3), hydrogen (H), and carbon monoxide (CO). However, only trace quantities of oxygen were present. At around 3.5 billion years, Earth's third atmosphere began to form as the first life forms- simple, unicellular bacteria- appeared.

122. Which is not a principle law of geology?
B. faulting
The principle laws of geology are:
- Principle of Uniformitarianism: Processes that are happening today also happened in the past.
- **Principle of Cross-Cutting Relations: A rock is younger than any rock it cuts across.**
- Principle of Original Horizontality: Rock units are originally laid down flat. Something happened to cause them to change orientation.
- Principle of Super Position: The rock on the bottom is older than the rock on top.
- Principle of Biologic Succession: Fossils correspond to particular periods of time.

123. The red beds are important because they indicate the presence of _____ in the geologic record.
C. Oxygen
Formation of Red Beds: The Animike Group- banded iron formations- form. These Red Beds are important because they herald the appearance of significant amounts of oxygen on the Earth. The red color is produced by rust. The rust indicates the presence of oxygen acting upon the ferrous material present in the ocean, and eventually, on the land. The presence of significant amounts of oxygen allows ozone to form, which in turn, screens out the harmful ultra-violet (UV) rays. This makes life possible outside of the protective confines of the ocean.

124. Tornadoes are most likely to occur in what season?
A. Spring
Tornado: an area of extreme low pressure, with rapidly rotating winds beneath a cumulonimbus cloud. Tornadoes are normally spawned from a Super Cell Thunderstorm. They can occur when very cold air and very warm air meet, usually in the Spring. Tornadoes represent the lowest pressure points on the Earth and move across the landscape at an average speed of 30 mph.

125. Which scale is used to measure hurricanes?
B. Saffir-Simpson Scale
The Fujita Scale is used to measure the intensity and damage associated with tornadoes. The Saffir-Simpson Scale is used to classify hurricanes into five categories, with increasing numbers corresponding to lower central pressures, greater wind speeds, and large storm surges. Richter Scale: the primary scale used by seismologists to measure the magnitude of the energy released in an earthquake.

www.ingramcontent.com/pod-product-compliance
Lightning Source LLC
Chambersburg PA
CBHW080537300426
44111CB00017B/2776